William Alexander Foster

Foster's first principles of chemistry

llustrated by a series of the most recently discovered and brilliant experiments.

Second Edition

William Alexander Foster

Foster's first principles of chemistry
Illustrated by a series of the most recently discovered and brilliant experiments. Second Edition

ISBN/EAN: 9783337221942

Printed in Europe, USA, Canada, Australia, Japan

Cover: Foto ©berggeist007 / pixelio.de

More available books at **www.hansebooks.com**

FOSTER'S

FIRST PRINCIPLES

OF

CHEMISTRY.

ILLUSTRATED BY

A SERIES OF THE MOST RECENTLY DISCOVERED AND
BRILLIANT EXPERIMENTS KNOWN TO
THE SCIENCE.

ADAPTED SPECIALLY FOR CLASSES.

BY W. FOSTER, A.M.,

PROFESSOR OF NATURAL SCIENCE, LEAVENWORTH COLLEGE, KANSAS.

SECOND EDITION.

NEW YORK:

HARPER & BROTHERS, PUBLISHERS,

FRANKLIN SQUARE.

PREFACE.

THE author of this strictly elementary series does not design it to take the place of the more elaborate works already in use, where extensive apparatus is at the command of the teacher ; but having been convinced from observation that the greater number of our academies and schools where Chemistry is professedly taught are destitute of the apparatus necessary to illustrate the text-books used, and that these text-books are not adapted to the wants of our primary schools, he respectfully presents this work, believing these objections to be in some measure obviated.

The instructor in this important branch of popular education will find a few dollars' worth of apparatus sufficient to enable him to perform all the most beautiful and striking experiments embraced in large treatises, as well as the newly-discovered ones contained in this.

The aim has been to divest the subject of technicalities, and to present each natural division in a strictly practical form, illustrated by diagrams and experiments so simplified as to be within the comprehension of the youth as well as the adult.

The Imponderable Agents have not been introduced,

for two reasons : 1st. The limits of this work would not permit it; 2d. They are invariably introduced and treated at length by authors upon Natural Philosophy.

The teacher and pupil will soon discern that this work is practically experimental, each of the numerous experiments contained in it having been performed by the author in various ways, and the simplest mode only described.

Every experienced instructor is aware of the difficulty he invariably encounters when attempting to illustrate a chemical decomposition to students who are not familiar with *Chemical Symbols*. With a view to imprint these more firmly upon the mind, they are invariably used (after once introduced) in place of the name of the element or compound for which they stand, the name following in parentheses. The teacher will also notice that a great number of new and simple diagrams have been introduced, in order to render decompositions and combinations entirely clear to the student. These diagrams should be drawn upon the black-board and explained, which will be found the most expeditious and thorough method of illustrating the chemical changes contained in each lesson. Thus the student will soon become fai .iliar with the *expressive language* of Chemistry, and necessarily delighted with the certainty and beauty of the results produced.

CONTENTS.

CHEMICAL APPARATUS.

THE Illustration gives a general view of the Apparatus furnished to illustrate FOSTER's CHEMISTRY. It has been prepared under the direction of the Author, and contains all that is necessary in order to perform the experiments introduced. It consists of the following articles:

Pneumatic Trough.	Brass Bladder Piece.
India-rubber Gas Bag.	Brass Double Connector.
Ivory Mouth-piece.	Glass Stoppered Retorts (3).
Retort Stand.	Glass Stoppered Receiver.
Spirit Lamp.	Glass Funnel.
Sand Bath.	Glass Funnel Tube.
Lead Tray.	Glass Gas Flasks (3).
Hydrogen Pistol.	Transfer Bell-glass, with Cap.
Hessian Crucibles.	Glass Test Tubes (13), with
Brass Stop-cocks (3).	Stand.
Brass Jets (2).	Glass Tubes, bent and strait.
Brass Tube for Bubble Pipe.	India-rubber Connectors.

The following Chemicals not usually to be obtained from country druggists, are also included:

Potassium.	Sodium.
Chlorate of Potassa.	Phosphorus.
Oxide of Manganese.	Fluor Spar.

The Apparatus is furnished, carefully packed for transportation, Price $23 00.

The Author has also prepared a large Engraved Chart of the Organic Elements, size 69 by 70 inches. It is beautifully colored and mounted on rollers, and will be found very useful in the class-room. Price $4 00.

FIRST PRINCIPLES OF CHEMISTRY.

LESSON I.

What does chemistry teach us?

CHEMISTRY teaches us how many kinds of matter there are in nature around us.

What does it also show?

It also shows us how one kind of matter behaves toward another when brought in contact.

Give the example.

Example: Lard and water will not mix, though we apply heat.

If lye be added, what follows?

Now if we add lye, all three bodies unite, and form soap.

What, then, do we infer?

So we infer that water does not attract lard, nor lard water; but that lye attracts both, and thus draws them together.

By what is the behavior of the lard, water, and lye controlled?

The behavior of the lard, water, and lye is controlled by what chemists call *Affinity.*

How are chemical force and chemical attraction used?

ty. Chemical Force and Chemical Attraction are used in the same sense.

Simple Affinity.

What is the color of the flame of a friction match?

Experiment 1. Ignite a friction match in the air.

What is its odor?

What has the free oxygen of the air united with?

It burns with a blue flame, emitting a suffocating odor. The *free* oxygen of the air has united with the *free* sulphur upon the match, and formed a poisonous gas.

What is the force called which united these two free bodies?

The force which united these two free bodies is called *Simple Affinity.*

Single Elective Affinity.

Give a definite explanation of Experiment. 2.

Exp. 2. Dissolve half an ounce of sugar of lead (*acetate of lead*), composed of acetic acid and lead,* in four ounces of water. Add a few drops of acetic acid to clarify the solution, and pour off the clear liquid into a wide-mouthed bottle. Suspend in it by a thread a slip of sheet zinc three fourths of an inch wide and two inches long.

In the course of an hour, what takes place?

In the course of one hour, brilliant foliated crystals of metallic lead will be seen shooting out from the zinc in every direction.

What appearance do these crystals gradually assume? What is the tree called?

These crystals will gradually assume the appearance of an inverted tree, which is called *Arbor Saturni* (lead tree). The action of the zinc upon the solution will be readily seen by a glance at the following diagram:

Fig. 1.

Draw the diagram upon the blackboard, and explain the change.

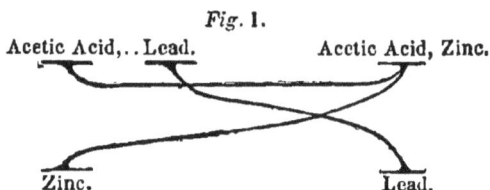

Acetic Acid,..Lead.   Acetic Acid, Zinc.

Zinc.   Lead.

The acetic acid and lead were held together by a certain force; the zinc had a

---

* Acetate of lead is really an acetate of oxide of lead, but the oxide in this experiment, as well as in several of the following, has been omitted for the sake of simplicity.

stronger attraction for the acetic acid than the lead had for it; hence the zinc united with the acetic acid, and the lead was set free in its metallic form.

Explain Exp. 3. *Exp.* 3. Fill a small test-tube one third full of the solution of sugar of lead, formed by *Exp.* 2. Now add a few drops of sulphuric acid. A white powder will be

Of what is the white powder composed? thrown down, which is composed of sulphuric acid and lead. At first the acetic acid and lead were united, but when the

What effect did the sulphuric acid have upon the solution of sugar of lead? sulphuric acid was brought in contact, it elected or chose the lead, and set the acetic acid free.

Fig. 2.

Acetic Acid, . . Lead.   Acetic Acid.

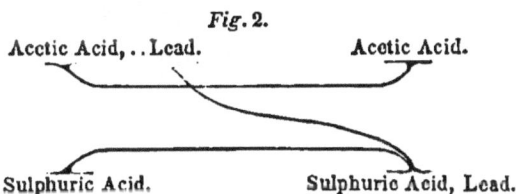

Explain the change by Fig. 2.

Sulphuric Acid.   Sulphuric Acid, Lead.

What is the common name of sulphate of iron? *Exp.* 4. Pour a dram or two of a transparent solution of sulphate of iron* (copperas) into a large test-tube, and add a

If iron be brought in contact with nutgall in solution, what color does the liquid assume? few drops of a colorless infusion of nutgall. A jet black is instantly imparted to these transparent liquids. Sulphate of iron is

What is sulphate of iron composed of? composed of sulphuric acid and iron. The

When the nutgall chooses the iron, what becomes of the acid? nutgall elects the iron and forms the ink, while the sulphuric acid is set free. A

What takes place when oxalic acid is added? few drops of a solution of oxalic acid will

* For the formation of solutions, see p. 134

How is the dark color destroyed?

destroy the dark color by removing the iron from the nutgall. Add a few drops of a solution of potassa, which will elect

How does potassa affect the solution?

the oxalic acid, when the iron will again unite with the nutgall, leaving the liquid black as before. Two or three drops of

What takes place when hydrochloric acid is added?

hydrochloric acid will combine with the potassa, taking it from the oxalic acid, which, now being liberated, unites with the iron, separating it from the nutgall, when the solution is again transparent. Add

When ammonia?

some liquid ammonia, which will combine with the hydrochloric acid, taking it from the potassa, which will now unite with the oxalic acid, taking it from the iron, which, being released again, combines with the nutgall, and the liquid assumes

If sulphuric acid be added, what changes follow?

a dark color. Lastly, add three or four drops of sulphuric acid, which will elect the ammonia, taking it from the hydrochloric acid, which now, set at liberty, unites with the potassa, taking it from the oxalic acid, which acid now combines with the iron, removing it from the nut-

Is the solution left black or transparent?

gall, when the whole is again transparent.

What is sulphate of copper commonly called?

Exp. 5. Into a weak solution of sulphate of copper (blue vitriol) drop some

What follows when liquid ammonia is added to a solution of sulphate of copper?

liquid ammonia. A brilliant blue color will be instantly imparted to the liquid, which is ammonia combined with copper.

Fig. 3.

Sulphate of copper=
Sulphuric Acid, Copper.

Sulphuric Acid
(set free).

Ammonia.

Ammonia, Copper.

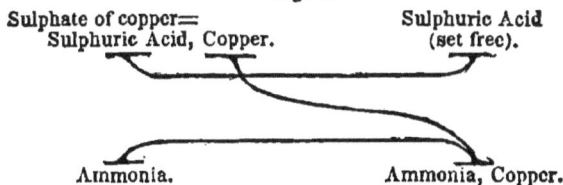

Draw and explain
Fig. 3.

The ammonia elects the copper, setting the sulphuric acid free.

If iron be immersed in a solution of sulphate of copper, what takes place?

Exp. 6. Into a strong solution of sulphate of copper dip the clean blade of a knife or an iron nail. Hold the metal in the solution for three or four minutes, when it will be covered with a beautiful coat of metallic copper.

Fig. 4.

Sulphuric Acid, Copper.

Sulphuric Acid, Iron.

Iron.

Copper (set free).

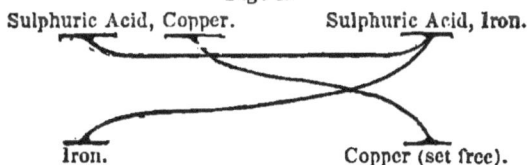

Draw and explain
Fig. 4.

Iron has a stronger attraction for sulphuric acid than copper has for it. Hence the iron elects the acid, and the copper is set free in its metallic form.

With what will a copper cent be coated if immersed in a solution of nitrate of mercury?

What does it then resemble?

Exp. 7. Immerse a clean copper cent in a solution of nitrate of mercury. It will soon be coated with mercury in its metallic form, when it will resemble a silver coin, and may be brightened with a piece of buckskin.

Fig. 5.

Nitrate of Mercury=
    Nitric Acid, Mercury.   Nitric Acid, Copper.

            Copper.   Mercury (set free).

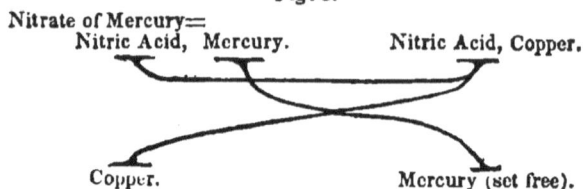

What is nitrate of
    mercury   com-
    posed of?
What did the cop-
    per elect?
Was the mercury
    still united with
    the nitric acid?

Give a general def-
    inition of single
    elective affinity.

The copper elected the nitric acid, and the mercury was set free and deposited upon it.

Single Elective Affinity, then, is that force which enables one body to elect or choose another which is combined with a third, which third body is always set free, while the first and second are united in a new compound.

LESSON II.

Double Elective Affinity.

What is produced
when   solutions
of alum and ace-
tate of lead are
poured together?

Exp. 8. INTO a solution of acetate of lead pour some solution of alum.* A white precipitate is the result.

Fig. 6.

Sulphuric Acid, Alumina.   Alumina, Acetic Acid.

Explain the figure.

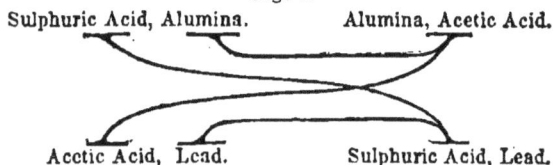

    Acetic Acid, Lead.   Sulphuric Acid, Lead.

* Alum is a double salt, composed of sulphate of alumina and sulphate of potassa. But the sulphate of potassa is inert in the decomposition.

Is the decomposition single?
Is but one new compound formed?
What becomes of the acids of the two primary compounds?
What has the sulphuric acid of the sulphate of alumina elected?
What has the acetic acid chosen?
Is a compound formed of sulphuric acid and lead soluble in water?

Are acetic acid and alumina, when united, soluble in water?

If solutions of bichromate of potassa and acetate of lead are poured together, what follows?
What is its common name?
Its chemical name?

Here we have a double decomposition, giving rise to two new compounds, acetate of alumina and sulphate of lead. The acids of the two primary compounds have changed places. The sulphuric acid of the sulphate of alumina has elected the lead of the acetate of lead, and the acetic acid has chosen the alumina. Sulphuric acid and lead, as has been shown, *Exp.* 3, form a compound which can not be dissolved in water, therefore it is visible. But acetic acid and alumina form a compound that is soluble in water, and hence it remains invisible in the solution.

Exp. 9. Instead of the foregoing, use solutions of bichromate of potassa and acetate of lead. A beautiful yellow powder will be formed, which is the common *chrome yellow* of the shops. Its chemical name is chromate of lead.

Fig. 7.

Chromic Acid, Potassa.   Potassa, Acetic Acid.

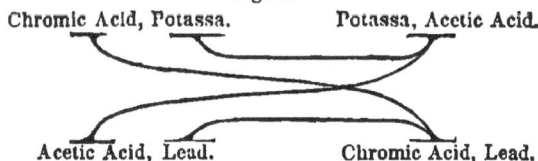

Acetic Acid, Lead.   Chromic Acid, Lead.

Explain Fig. 7.

What has the chromic acid combined with?
What then becomes of the acetic acid?

Define double elective affinity.

The chromic acid has united with the lead, and the acetic acid with the potassa.

Double Elective Affinity, as will be seen from *Exp.* 8, 9, is that force which

causes a decomposition of two compound
bodies, and a union of their parts to form
two new and distinct compounds.

*Further Experiments in Double Elect-
ive Affinity.*

Exp. 10. Use solutions of bichromate
of potassa and nitrate of silver. A rich
carmine color will pervade the liquid. The
two new compounds are nitric acid and

potassa (nitrate of potassa), and chromic
acid and silver (chromate of silver), which
is the brilliant precipitate.

Exp. 11. Into a solution of bichloride
of mercury pour some solution of iodide
of potassium (composed of iodine and potassium). From these perfectly transparent solutions a beautiful vermilion color is
obtained. The chlorine of the chloride of
mercury has united with the potassium of
the iodide of potassium, while the iodine
combined with the mercury, forming iodide of mercury, the vermilion precipitate.

Fig. 8.

Iodine, Potassium.   Potassium, Chlorine.

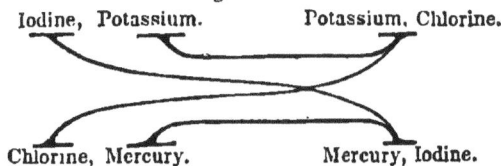

Chlorine, Mercury.   Mercury, Iodine.

From the foregoing experiments in Simple, Single Elective, and Double Elective

Affinity, we learn that chemical force acts only between particles of different kinds of matter at invisible distances; and that when bodies unite chemically, their properties are entirely changed. Hence the resulting compound differs materially from either of its constituents.

What changes has chemical affinity already effected in bodies?

We have already seen that chemical affinity effects remarkable changes in the *color* of bodies. We will now consider *change* of *form* and *change* of *temperature* as connected with this invisible force.

What is next to be considered?

LESSON III.

Change of Form.

If chalk be added to hydrochloric acid, what takes place?

Exp. 12. FILL a wine-glass half full of hydrochloric acid, and add small pieces of chalk until effervescence ceases. A portion of the solid chalk has passed off in the form of a gas, which caused the bubbling.

What power, then, has chemical affinity over solids?

Chemical affinity has power to convert solids to gases.

Exp. 13. Pour off some of the yellowish liquid formed by *Exp.* 12 into a test-tube, and add a few drops of strong sulphuric acid. The two liquids are converted into a beautiful white solid, which is

Explain Exp. 13.

What power has chemical affinity over liquids? sulphate of lime.　Chemical affinity some-times changes liquids to solids.

Exp. 14. Rinse out one wine-glass with hydrochloric acid, and another with liquid· ammonia.　Turn them bottom upward, and bring their mouths quickly together.

Explain Exp. 14. A white fume will fill both glasses, which is composed of small particles of a solid body (hydrochlorate of ammonia).　The two, liquids adhering to the sides of the glasses were first converted into gases, and from gases to the solid form.　Chem-How does affinity act on gases?ical affinity converts gases to solids.

Change of Temperature.

How may loaf-sugar be rapidly converted into charcoal? *Exp.* 15. Dissolve some loaf-sugar in the least quantity of warm water.　Let it remain ten minutes or more with some undissolved sugar in the bottom of the vessel.　It should be frequently stirred with a glass rod.　Fill a wine-glass half full of the solution, and add strong sul-phuric acid until the glass is nearly full. The sugar will be rapidly converted into What is sugar composed of?charcoal.　Sugar is composed of carbon What effect is produced by the acid?and water.　The acid unites with the water, and sets free a ˉsufficient amount What remains behind?of heat to evaporate it, while the carbon is left behind in the form of finely pulver-ized charcoal.　In this experiment we have, first, a change of temperature; second, a liquid converted to vapor; and third, a liquid converted to a solid.

What first takes place?

What second and third?

We have now learned that chemical affinity has power to change the form, temperature, and color of bodies. Indeed, it is this unseen force, directed by a Divine Hand, that produces that endless variety of colors which deck the world in which we live. Various materials in transparent solutions are absorbed into the structure of plants and animals, and there undergo wonderful changes, many of which may be successfully imitated in the laboratory. Chemical force, however, is sometimes directed with such matchless skill in the formation of colors that they can not be imitated by man. The delicate tints which we witness in the butterfly's wing, in the peacock's tail, and in the petals of flowers, have ever baffled the skill of the artistic chemist.

What have we now learned in relation to the invisible force called affinity?

How are colors in nature produced?

What are absorbed into the structure of plants and animals?

Can the changes be imitated in the laboratory?

What is said of the butterfly's wing and the petals of the flower?

Chemical Affinity is affected by the *Imponderable Agents*. Light, Heat, Electricity, and Galvanism, are called imponderable agents, because they have no weight. Heat and Galvanism usually promote decomposition (*analysis*). Light and Electricity facilitate combination (*synthesis*). But, as these agents belong

What agents affect chemical affinity?

Why are they called imponderable?

What do heat and galvanism promote?

What do light and electricity promote?

To what branch of natural science do the imponderable agents more properly belong? more properly to Natural Philosophy, they will not be treated of in this work.

LESSON IV.

Chemical Elements.

Bodies that chemists have been unable to decompose are called what?
Give an example.
How many elements have already been discovered?
Which are the most important of the elements? SUCH bodies as chemists have not been able to decompose are called *Elements*, or simple bodies; as, Iron, Sulphur, Nitrogen, &c. Of these elements, sixty-five have already been discovered. The most important of the elements are such as combine to form the structure of our bodies, those of the lower animals, and plants.

Name them. They are sixteen in number: Hydrogen, Oxygen, Chlorine, Nitrogen, Carbon, Sulphur, Phosphorus, Silicon, Fluorine, Calcium, Aluminum, Sodium, Potassium, Magnesium, Manganese, and Iron.

Chemical Symbols.

How do chemists now represent elementary bodies?
Why? For the sake of convenience, chemists now represent elementary bodies by symbols, thus:

What does H stand for? O?
Cl? N? C?

Hydrogen is represented by H.

Oxygen   "   "   O.

Chlorine   "   "   Cl.

Nitrogen   "   "   N.

Carbon   "   "   C.

S! P! Si! F! Ca!	Sulphur is represented by S.
	Phosphorus " " P.
	Silicon " " Si.
	Fluorine " " F.
	Calcium " " Ca.
Al! Na! What is its Latin name?	Aluminum " " Al.
	Sodium (Latin *Natrum*) Na.
K! Its Latin name?	Potassium (Latin *Kalium*) K.
	Magnesium " " Mg.
Mg! Mn?	Manganese " " Mn.
Fe? The Latin?	Iron (Latin *Ferrum*) " Fe.

What do these sixteen simple bodies form? These sixteen simple bodies, in various combinations, form rocks and soils, as well as all organic bodies upon the surface of the earth, together with the atmosphere and the waters of the ocean.*

LESSON V.

Division of Matter—Atoms.

What does the word atom mean? What are all bodies composed of? What is the probable shape of atoms? Have atoms ever been seen? Why not? Is their existence established?

THE word *Atom* means that *which can not be divided.* All bodies are composed of atoms, which are probably globular. They have never been seen, as they are so small as to baffle the powers of the microscope. Their existence, however, is well established.

* Other elements are found in very minute quantities in rocks, soils, and organic bodies.

B

Atomic Numbers.

What is found by experiment?

It is found by experiment that simple bodies unite only in fixed quantities. For

Of what is water composed?

example, the water we drink is composed of two gases, Oxygen and Hydrogen. It

How much, by weight, of each?

requires just one ounce of H to unite with eight ounces of O, and if we attempt to combine more or less of either of these

Why must the proportion be invariably preserved?

elements, we shall fail. The proportion must invariably be preserved, or a residue of either the one or the other element will remain uncombined.

How do all elementary bodies unite?

All elementary bodies unite with each other in determinate quantities by weight.

Why is hydrogen taken as one?

H (hydrogen), being the lightest of the elements, is taken as unity, and all other elements are compared with it. Thus,

How many grains of hydrogen unite with thirty-five grains of chlorine? Will H unite with calcium? Will O?

1 grain of H will unite with 35 grains of Cl (chlorine), 6 of C (carbon), 8 of O (oxygen), &c. But H (hydrogen) will not unite with Ca (calcium), and O will. Now we find that 20 grains of Ca will unite with 8 grains of O, and 8 grains of

How is H indirectly compared with calcium? When is this indirect mode of comparison made? When we speak of H, what is always associated in the mind? of O? of C?

O with 1 of H. Thus Ca is indirectly compared with H. This indirect mode of comparison is always made where the element does not unite with H. When we speak of H, 1 is always associated in the mind; of O, 8, and of C (carbon), 6, &c.

What are these atomic numbers sometimes called?
These atomic numbers are sometimes called *Definite Proportions, Chemical Equivalents,* and *Combining Numbers.*

What does an atomic number express?
These different expressions, however, mean the same thing, namely, the quantity by weight of one element that is required to satisfy its affinity for 1 of H or 8 of O. The following table exhibits the atomic number of each of the sixteen organic elements.

What is the atomic number of H (hydrogen)?
Of O (oxygen)?
Of Cl (chlorine)?
Of N (nitrogen)?
Of C (carbon)?
Of S (sulphur)?
Of P (phosphorus)?
Of Si (silicon)?
Of F (fluorine)?
Of Ca (calcium)?
Of Al (aluminum)?
Of Na (sodium)?
Of K (potassium)?
Of Mg (magnesium)?
Of Mn (manganese)?
Of Fe (iron)?

$H = 1.$*	$F = 19.$
$O = 8.$	$Ca = 20.$
$Cl = 35.$†	$Al = 14.$
$N = 14.$	$Na = 23.$
$C = 6.$	$K = 39.$
$S = 16.$	$Mg = 13.$
$P = 32.$	$Mn = 28.$
$Si = 21.$	$Fe = 28.$

* On the Continent of Europe the atomic number of oxygen is taken as 100, in which case hydrogen would be 12·5, carbon 75, &c.

† The atomic number of some of the elements is composed of a whole number and a fraction. The fraction, for the sake of simplicity, has been omitted where it was less than one half, and a unit added where it was greater.

LESSON VI.

Multiple Proportions.

When elements unite in more proportions than one, is the quantity variable? What is said of the greater?

WHEN elements combine in more proportions than one, the quantity of each is also fixed, but the greater is always a multiple of the less by a whole number : *e. g.*,

Give the example.

1 at. of N(14) unites with 1 at. of O(8).

1 " N(14) " " 2 ats. of O(16).

1 " N(14) " " 3 " O(24).

1 " N(14) " " 4 " O(32).

1 " N(14) " " 5 " O(40).

Or, in strictly chemical language,

Give it in chemical language. What are 2, 3, 4, and 5 multiples of?

$NO, NO^2, NO^3, NO^4, NO^5$.

16, 24, 32, and 40 are multiples of 8 by the whole numbers 2, 3, 4, and 5. Some-

How do elements sometimes unite? Can an atom be divided? What, then, is presumed?

times elements unite in the proportion of 1 to $1\frac{1}{2}$; but as an atom is not divisible, it is presumed that 2 of the one is combined with 3 of the other, which will pre-

What example is given?

serve the proportion : thus, $1 : 1\frac{1}{2} :: 2 : 3$; *e. g.*, 2 atoms of iron unite with 3 atoms

When elements unite in the proportion of 1 to $1\frac{1}{2}$, what term is applied to them?

of oxygen to form iron rust. When elements unite in the proportion of 1 to $1\frac{1}{2}$, the term *sesqui-* is applied to them. Hence iron rust is a sesquioxide of iron.

Per- or Hyper-, -ic, -ous, and Hypo-.

When one element unites with another,
the prefix *per* or *hyper* is used to denote
the greater though indefinite quantity of
the first element named, except when applied to O acids, when it implies the greater amount of O only. The termination
-ic in O acids indicates more O than *-ous*,
and *-ous* more than *hypo-* : e. g.,

Cl and O (ClO) form *hypo*chlorous acid.

Cl " O^3 (ClO^3) " chlorous "

Cl " O^4 (ClO^4) " *hypo*chlor*ic* "

Cl " O^5 (ClO^5) " chloric "

Cl " O^7 (ClO^7) " *per* or *hyper*chloric acid.

How are per or hyper used?

When applied to oxygen acids, to what do these two prefixes apply?

What does the termination ic indicate?

What ous?

What hypo?

Give the example.

Proto-, Deuto- or Bi-, Trito- or Ter-, and Quadro-.

Proto- denotes 1 atom of each of two
bodies which are united in a compound ;
*proto*xide of hydrogen (water) is composed
of 1 atom or equivalent of H and 1 of O.
Deuto- or *bi-* is used where 2 atoms of
one body are combined with 1 atom of
another ; as, *deut*oxide or *bin*oxide of nitrogen (NO^2). *Trito-* or *ter-* is prefixed
where 3, and *quadro-* where 4 atoms of
one body are united with 1 atom of another. *Quadro* is seldom used.

What does proto denote?

Deuto or bi?

Trito or ter?

Quadro?

Is quadro often used?

LESSON VII.

-ide, -uret, -ate, and -ite.

How is the termination ide applied?

O (oxygen), and all elements ending in *-ine*, when united with another, take the termination *-ide* when the compound is

Give examples.

not acid; as, ox*ide* of calcium (lime), chlor*ide* of sodium (common salt). The O, however, in a few instances, is repre-

What is said of a? Give examples.

sented by *-a;* as, sod*a* for oxide of sodium, potass*a* for oxide of potassium, and

What of other non-metallic elements?

silic*a* for oxide of silicon. Other non-metallic elements have the termination

Give examples.

*-uret;** sulph*uret* of iron, carb*uret* of hydrogen, &c.

When the acid ends in ic, how does the salt terminate?

Salts formed from an acid and a base end in *-ate* when the acid ends in *-ic*; sulphur*ic* acid and soda form sulph*ate* of

If the acid ends in ous, how?

soda. When the acid ends in *-ous*, the salt ends in *-ite.* Sulphur*ous* acid and soda form sulph*ite* of soda. If the acid

When hypo is prefixed to the acid, what of the salt?

has the prefix *hypo-*, the resulting salt always retains it; *hypo*sulphur*ous* acid and

Give an example.

soda form hyposulph*ite* of soda.

---

* Professor Gregory, of Edinburgh, and some other authors, use the termination *-ide* in preference to

Formulas.

What is a chemical formula?
What is the formula for water?
When more than one atom of an element is to be expressed, how is it accomplished?
Write an example upon the blackboard.

How do compound bodies have their symbols separated?
Write an example.

What does a large figure accomplish when placed before a symbol or formula?
When placed before brackets, what?

Two or more symbols placed together constitute a formula. HO is the formula for water. When more than one atom of an element is to be expressed, it is accomplished by placing a small figure either above or below the symbol, thus: SO^2 or SO_2 shows that one atom of sulphur is combined with two atoms of oxygen.

Two or more compound bodies united chemically have their symbols separated by a , or the sign $+$; SO^3, HO, or SO^3 $+$ HO, indicates that the compound SO^3 is chemically united* with the compound HO. When a large figure precedes a formula or a symbol, it multiplies all the symbols until a , or the sign $+$ intervenes. A large figure placed before brackets multiplies all the symbols included between them.

-uret. But as there is more euphony in sulphuret than in sulphide, chemists in the United States have adopted the termination -uret instead of -ide.

* Many bodies are mechanically mixed, though not chemically united. When bodies are chemically united, the compound invariably possesses different properties from either of its component parts. But if mechanically mixed, no change of properties will have taken place.

LESSON VIII.

Hydrogen. Equivalent 1. *Symbol H.*

Explain the appa-
ratus for obtain-
ing H.
Exp. 16. Introduce a handful of slips of zinc into a bottle having a wide mouth, and containing at least a pint. Add a half pint of water, and insert a cork, which should be nicely fitted with a glass funnel tube reaching nearly to the bottom of the bottle; another tube, bent at right angles,* should merely pass through the cork. Connected with this latter tube there should be one of the same size, and in shape resembling the letter S, joined by means of an india rubber connector. Now pour an ounce of SO^3, HO (sulphuric acid) through the funnel-tube, and H (hydrogen) will be rapidly disengaged, How may H be collected for experiment? which may be collected in half gallon candy-jars over water for experiment, by means of the pneumatic trough (Fig. 9). The trough should be filled with water, and, in order to remove the air from the

* Glass tubes may be readily bent by holding them in the flame of a spirit-lamp and applying a gentle pressure.

Fig. 9.

How must the air be removed from the jar previous to admitting the gas? jar previous to admitting the gas, it should be dipped into the trough, and when filled with water, inverted upon the pneumatic shelf, beneath the surface. Reject the

Why are the first portions of the gas to be rejected? first portions of gas generated, as they are mixed with the air which was in the generator. Now pass the tube which conducts the gas under the mouth of the jar,

Why does H displace the water in the jar? which will soon be filled with H, the water being displaced by the gas. As soon as H commences to escape from the

How may the jar be removed when filled with H? mouth of the jar, it may be removed by sliding it upon a common dinner-plate filled with water, taking care to keep the mouth of the jar beneath the surface. Other jars may be filled in a similar manner. Remove the external tube, and con-

Explain the wash-bottle. nect the wash-bottle (Fig 10), which,

B 2

Fig. 10.

like the generator, consists of a bottle and

What should it contain?

two tubes. It should contain a solution of potassa (one half drachm of potassa to three ounces of water), and the end of the

Why should the end of the tube be tied over with flannel?

conducting tube should be tied over with a piece of canton flannel, in order to force the gas into the solution in small bubbles. The solution should not cover the mouth of the tube more than half an inch deep. Press out the air from a gas-bag containing about one gallon, and connect it with the washer (wash-bottle). It will soon be filled with H, sufficiently pure for practi-

What object is accomplished by the washer?

cal purposes. The object of the washer is to remove small particles of SO^3HO (sulphuric acid), which always pass over mechanically mixed with the gas. The

What materials were used?

materials used for obtaining H were HO

(water), SO^3HO (sulphuric acid), and Zn. (zinc). Fig. 11 will present a clear view of the decomposition.

Fig. 11.

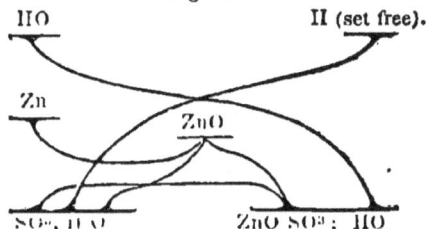

[A semicolon separating two symbols shows that they are not chemically united.]

What does an atom of zinc elect?
What does it elect oxygen from?
What does zinc united with O form?
What is set free?

An atom of Zn elects an atom of O from the HO, which was chemically united with the SO^3, and forms ZnO (oxide of zinc), while one atom of H is set free and passes upward in its gaseous form. The action would here cease, as the surface of the metal is now covered

What does an atom of liberated SO^3 unite with?

What salt is formed?

with the oxide, but an atom of liberated SO^3 combines instantly with this oxide, and dissolves it, forming sulphate of oxide of zinc. This sulphate would again

Why does not this salt again put an end to the decomposition?

put an end to the action of the metal, as it would cover the surface, but the free HO (water) dissolves it as fast as it is formed. Thus a constant clean metallic surface is presented to the HO of the SO^3,

How long will H be liberated?

and a constant liberation of H follows until some of the materials are exhausted.*

* The commonly received theory of the formation

LESSON IX.

Hydrogen (continued).

What takes place when hydrochloric acid is used instead of sulphuric?

Exp. 17. Use HCl (hydrochloric acid) instead of SO³HO, and the process will be

of H is, that the zinc unites with the O of the *free* water, instead of the water which is combined with the SO³, which conclusion resulted probably from the well-known fact that several of the metals, when presented to water without the presence of an acid, will slowly become oxydized. Hence the SO³HO (sulphuric acid) was supposed to act as a solvent of this oxide, while the *free* water was decomposed. This theory, however, is erroneous, as will appear from the following considerations : 1st, it is a fixed law, to which there are no exceptions, that when a metal is presented to two compounds containing O, and it elects this element from one of them, it invariably takes it from the compound which is easiest decomposed. Ternary compounds are easier decomposed than binary. SO³HO is a ternary compound, and *free* HO is binary. Hence the zinc must take Q from the SO³HO. 2d, Sulphuric acid, in its active state, is SO⁴, H, and here the force which holds the H and O together is divided into three parts instead of two, in the case of water ; therefore *free* HO requires more force to decompose it than HO when chemically combined with S. The metal, it is true, will slowly elect O from free water when the acid is not present, but as soon as the acid is presented, the action is entirely changed. as O is now presented to the metal which it

more simple. In this decomposition the

Fig. 12.

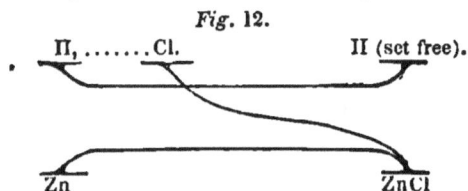

H,.......Cl.　　　　　　H (set free).

Zn　　　　　　ZnCl

Draw and explain
Fig. 12.

zinc simply unites with the Cl of the

Give the formula. acid, and the H is liberated. Formula,
HCl; $Zn = ZnCl$; H.

Exp. 18. Attach a bubble-pipe to the
gas-bag containing H, and fill soap-bubbles with the gas; they rapidly ascend.

Is H heavier than
air? Hence H is lighter than air. Apply a
lighted candle to a bubble after it is disengaged from the pipe. It will burn. H

Is it combustible? is a combustible gas.

What experi-
ments prove that
H is lighter than
air, and that it is
combustible? *Exp.* 19. Lift a jar of H from the plate;
keep its mouth downward, and quickly
pass a lighted candle upward into it; a

requires but a comparatively slight force to separate
from its previous combination. 3d, The quantity of
H obtained is dependent upon the quantity of acid
added instead of the quantity of free water. We may
safely draw the conclusion, then, that the free water
is not decomposed, but simply acts as a solvent of the
ZnO, SO^3 (sulphate of oxide of zinc); that salt, not
being soluble in the acid, would remain as a solid incrustation upon the surface of the metal, if water were
not present to dissolve it, and thus the decomposition
of the acid would cease. If sulphate of oxide of zinc
were soluble in sulphuric acid, the gas would be readily disengaged without the addition of water.

When H burns, what is the color of the flame? What experiment proves this?

slight explosion will take place, and the gas will burn with a blue flame at the mouth of the jar, but the candle will be extinguished while immersed in it. Slowly withdraw the candle downward, and it will be relighted by the burning gas. Thus the candle may be extinguished and relighted for several times in the same jar.

Will H allow any body to burn in it?

H will burn itself, but will allow no combustible body to burn in it. " It is combustible, but a non-supporter of combustion."

What is Exp. 20?

Exp. 20. Remove a jar of H from the water, and place it upon the table with the mouth upward ; apply a lighted match to the mouth of the jar as quickly as possi-

Will water extinguish the flame?

ble. The gas is set on fire. Pour in water ; the flame is not extinguished.

Exp. 21. Attach a jet or pipe-stem to the H washer, as the gas is set free from the generator. Wait four or five minutes for the gas to expel the air, and apply a lighted match to the jet. The gas burns. Drop some iron filings into the flame. They burn vividly. Hold a small, dry

Is much heat produced when H burns? How would you prove this? What is formed when H burns in the air?

bottle over the flame, and it will soon be filled with watery vapor. When H burns in air, an intense heat is produced, and the resulting compound is water (HO).

Explain Exp. 22.

Exp. 22. Place the broken beak of a

retort over the ignited jet of H (a large
glass tube is better). Pass it downward
(see Fig. 13) until a penetrating tone is

Fig. 13.

What is the tone called? produced. The tone is not the science of
music, but the *music of science*. It is
What is its cause? probably produced by a series of slight ex-
plosions, which are unheard unless partly
confined.

Explain Fig. 14. *Exp.* 23. Prepare a leaden cannon or
tin tube, Fig. 14, holding about three gills,

Fig 14.

with an orifice near the breech one tenth

of an inch in diameter. Fit a long cork to the muzzle air-tight. It must pass into the tube at least one inch. Now place the thumb of the left hand firmly upon the orifice, and hold the muzzle over a jet of H (not ignited) until about one third of the air contained in the tube or cannon has been displaced by the gas. Keep its mouth downward, with the thumb firmly pressed upon the orifice, and force in the cork by placing it upon the table and pressing the tube down over it. Have a slip of paper folded several times lying near a lighted candle. Ignite one end of it, and, having laid the tube upon the table, remove the thumb, and quickly apply the flame to the orifice. A loud report will follow. Hence two volumes of air and one of H form an explosive mixture.

How much of the air should be displaced by the H?

To what is the explosion due?

Exp. 24. Hold a two-ounce phial over a jet of H (not ignited) for a short time. Turn the mouth upward and apply a flame. A whizzing report will follow, which is owing to the successive union of the particles of H with the particles of O of the air.

What causes the whizzing report?

Synopsis of H.

What is said of the weight of H?

It is the lightest of all known elements, on account of which property it has been

Why has it been used for filling balloons. It is a gas which,
used for filling
balloons?
Has it color? when pure, is without color, taste, or
Taste? Smell?
What imparts an smell. As it is ordinarily formed, it has
odor to it as it is
commonly form- an unpleasant odor, which arises from the
ed?
United with O, impurities it contains. It unites with O
what does it
form? to form HO (water). H is combustible,
Is it combustible?
Will it support but will not support combustion, and
combustion?
How does it form forms, with air, an explosive mixture. It
an explosive
mixture? is not poisonous when breathed, if it is
Is it poisonous
when breathed? pure, though it has a peculiar effect upon
What effect has it
upon the voice? the voice, a shrill, squeaking tone being

Why would an imparted to it. An animal placed in H
animal placed in
H die? would soon die for want of O.

LESSON X.

What is the
equivalent of O? *Oxygen. Equivalent* 8. *Symbol* O.
Its symbol?

What mixture is $Exp.$ 25. Pulverize three parts of
used for obtain-
ing O? ClO^5, KO (chlorate of potassa) with one
part of MnO^2 (peroxide of manganese),
and introduce three or four ounces of the
Into what is the mixture into a glass retort containing a
mixture placed?
pint. The retort should be supported by
a stand, Fig. 15. Place the retort in a
How is heat ap-
plied? sand-bath, and apply the strong heat of
What is first driv- a spirit-lamp, which will soon drive over
en over?
the air contained in the retort, and a rap-
What follows? id disengagement of O (oxygen) will soon

Fig. 15.

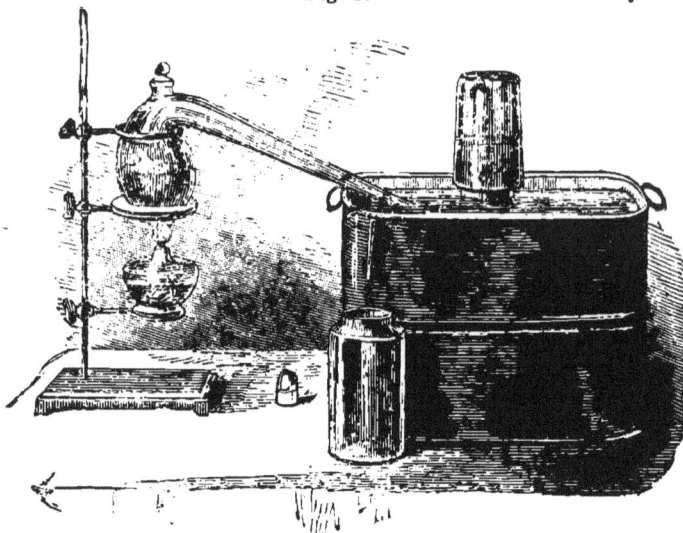

How may O be collected?

follow, which may be collected over water in half-gallon candy-jars, as in Exp. 16.* Having obtained the quantity required, remove the beak of the retort from the

Why should the beak of the retort be removed from the water before the heat is lessened?

water, for if the heat be first removed, the gas in the retort will condense, and the water will flow in upon the hot glass, when it will break in pieces. When the

How may the remaining materials be removed from the retort?

retort has cooled, the remaining materials may be washed out with water.

What were the substances used for obtaining O?

The substances used for obtaining O were chlorate of potassa and peroxide of

* The sand-bath consists of a tin, sheet iron, or copper vessel, nearly the shape of the bottom of the retort, filled with dry sand. Its object is to diffuse the heat equally over the surface of the glass, and thus lessen the liability of its breaking.

Was the peroxide of manganese decomposed?

How did it act?

When a body assists to decompose another without itself undergoing a change, what is the action called? Can we decompose chlorate of potassa without the presence of the manganese?

Is any more heat required?

manganese. The peroxide of manganese, however, has not been decomposed, but simply acted upon the chlorate of potassa by its presence. Where one body assists to decompose another without itself undergoing a change, the action is called *presence* or *catalytic action*. The chlorate of potassa may be decomposed and its O set free without the presence of the manganese, but the heat required would be about twice as great.

Fig. 16.

Draw the diagram upon the blackboard.

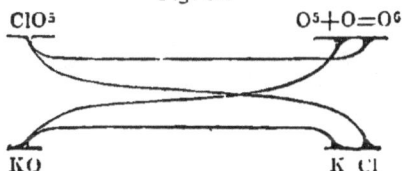

ClO^3 $O^5 + O = O^6$

KO   K Cl

Of what is chloric acid composed?
What is the composition of potassa?
What takes place when heat is applied?

Chloric acid is composed of Cl and O^5. Potassa is composed of one atom of the metal K and one of O. When heat is applied to this salt, the Cl unites with the K, setting free five atoms of O from the chloric acid and one atom from the potassa, while the new compound, KCl (chloride of potassium), remains in the retort,

What new compound remains in the retort?

With what is it mechanically mixed?

mixed mechanically with the oxide of manganese (MnO^2).

How may a jar of O be transferred from the plate to the table for Exp. 26?

Exp. 26. Remove a jar of O from the plate by placing the mouth of it beneath the surface of the water and allowing the plate to sink. Now slide a piece of tin or glass wrapped with newspaper under the mouth of the jar, which being held with one hand, the jar may be removed to the table and placed in an upright position.

If a living mouse be placed in O, what do his actions show? Introduce a living mouse into O, and immediately cover the jar. His actions will soon show that he lives faster than he did How will he breathe? in common air. He will breathe more rapidly, and often play " fantastic tricks." What will follow? Symptoms of stupor follow these antics, and if not quickly removed from the jar, death will ensue. The animal has lived too fast.

What two facts do we learn by this experiment? In this experiment we not only learn that O supports animal life, but that, when breathed pure, animals live too fast. With what has the Creator diluted this life-sustaining substance? Hence a benevolent Creator has diluted this life-sustaining substance with four times its bulk of a gas which neither supports nor destroys, but which is entirely What is the gas called? passive when breathed. It is called *Nitrogen*.

Exp. 27. Place a piece of roll sulphur

Fig. 17.

of the size of a pea upon a capsule, which should be two and a half inches high, having a cup of copper. (See Fig. 17.) Set the capsule upon a dinnerplate filled with water, and ignite the sulphur. Notice how

How does sulphur burn in air?

it burns in air. Now place over it a jar of O, the mouth of which may rest upon the plate beneath the surface of the water.

How in O?

The sulphur will burn with increased

With what will the jar be filled?

brilliancy, and the whole jar will be filled

With what has the sulphur united?

with a beautiful blue fume. S (sulphur)

In what proportion?

has united with O in the proportion of 1

What is the name of the resulting compound?

to 2. Hence the resulting compound is SO_2 (sulphurous acid).

Exp. 28. Cover a jar of O, as in Exp. 26, and place it upon the table. Attach a piece of tallow-candle to a wire not less than one foot long. Light the candle, and allow it to burn for a short time.

Explain Exp. 28 in full.

Extinguish the flame, and introduce it, with the wick glowing, into the jar of O. It will burst into vivid combustion with a slight report. It may be extinguished and relighted in this way for several times in the same jar of O.

What is the result of the combustion?

The result of the combustion is CO_2 (carbonic acid) and HO (water).

If a piece of the bark of charcoal be ignited and plunged into a jar of O, what beautiful effect is produced?

Exp. 29. Instead of the candle, attach a piece of the bark of charcoal to a wire; ignite one corner of it, and plunge it into a jar of O arranged as in Exp. 28. The jar must be kept covered, and the burning coal must not come in contact with its side. Brilliant stellated scintillations will burst from the coal. C (carbon) has

What has united with O?

In what propor-
tion?
What is the chem-
ical name of the
resulting com-
pound?
How much phos-
phorus may be
used in place of
the sulphur?
What of the light?
With what is the
jar filled?
What compound
is formed?

united with two atoms of O, forming CO^2 (carbonic acid).

Exp. 30. Instead of the sulphur in Exp. 27, use a piece of P (phosphorus) half the size. The light is very intense, and when it ceases the jar will be filled with a white fume, which is PO^5 (phosphoric acid), the result of the combustion.

How should a
watch-spring be
prepared for
burning in O?

Exp. 31. Hold one end of a broken watch-spring in the flame of a spirit-lamp until the temper is removed. Now file it to a thin, sharp edge. Cut off the upper end of a friction match so as to leave a clean pine stick a little more than half an inch long. Force the sharp edge of the metal into the stick nearly a quarter of an inch, and it will be ready for use. Have

How is the jar of
gas to be ar-
ranged?

a half-gallon candy-jar three quarters filled with O, the remaining space being occupied with water. Transfer this to the table as in Exp. 26. Ignite the end of the pine stick and immerse it dexterous-

Why should it be
kept covered?

ly into the jar, which should be kept covered in order to exclude the air. The

How does the pine
stick burn in O?

stick burns rapidly until it reaches the fuzzy edge of the metal, which now takes

What appearance
is presented as
the steel takes
fire?

fire, and throws out coruscations of light of surpassing beauty. Hence we learn

What do we learn
from the forego-
ing experiments
in O?

that O is the great supporter of combustion, and also the great sustainer of animal

What is O in its elementary form? life. In its elementary form it is a gas having neither taste, smell, nor color. It

Is it combustible? Do combustible bodies burn in it? is not combustible, though combustible bodies burn in it with great brilliancy.

How are all ordinary combustions produced? All ordinary combustions are produced by the direct union of O with the combusti-

Of what is tallow chiefly composed? ble body. Tallow is chiefly composed of C (carbon) and H (hydrogen). When a

When a candle burns in air, with what do carbon and hydrogen unite? candle burns, C and H chemically unite with the O of the air, forming two compounds, CO_2 (carbonic acid) and HO (water). H burns, as has already been seen,

What is the color of the hydrogen flame? Does the O of the air first unite with the H? Why is this? Exp. 19, page 37, with a blue flame; and as the O of the air has a stronger affinity for H than for C, the H

What causes the blue appearance of the lower portion of the flame of a candle? burns first, which accounts for the blue appearance, *a a*, Fig. 18, of the lower portion of

In what form does the carbon pass upward? the flame. The C, now freed from the H, passes upward in the form of a dark cone, *b*, and, being

With what does it now unite? What is the color of the flame produced by the carbon? What surrounds the outer portion of the flame? What is the cause of this? intensely heated, combines with O and burns with a yellow light, *c*. A pale envelope, *d*, surrounds the outer portion of the flame, which is caused by an incomplete supply of C.

Fig. 18.

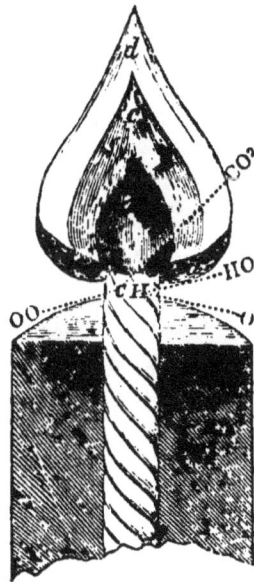

When O unites with a combustible body, heat is liberated, the amount of which depends upon the quantity of O consumed. The smith uses the bellows in order to bring an additional supply of O in contact with the charcoal (charcoal is nearly pure carbon).

O is the most widely diffused of the elements, constituting eight ninths of all the waters upon the surface of the earth, and about one fifth of the atmosphere. It also enters largely into combination with other elements to form rocks and soils.

Marginal notes:

When O unites with a combustible body, what is liberated?

Upon what does the quantity of heat depend?

How does the smith bring a large supply of O in contact with the charcoal?

What is charcoal?

What is said of the diffusion of O?

What part of water is composed of it?

What of the air?

Of rocks and soils?

LESSON XI.

Protoxide of Hydrogen (Water).
Equivalent* 9.   Symbol HO.

Exp. 32. Fill an India-rubber gas-bag with O by means of the bell-glass and pneumatic trough. Fill the bell-glass with the gas, with the stop-cock closed. Press the air from the gas-bag, and attach it to the stop-cock. Now open the passage from the bell-glass to the gas-bag by

Marginal notes:

What is the chemical name of water?

Its equivalent? Symbol?

How may a gas-bag be filled with O?

Describe the process in full.

What is the equivalent of a compound?

* The equivalent of a compound is the sum of the equivalents of its elements. The equivalent of O is 8, and that of H is 1; $8+1=9$, the equivalent of the compound protoxide of hydrogen.

turning the stop-cock, and press the bell-glass into the water. The superior weight of the water will force the gas upward into the gas-bag. (See Fig. 19.) The

Fig. 19.

Upon what is the gas-bag to be placed when filled? stop-cock may now be closed, and the gas-bag placed upon a block. Attach the jet, which should be elevated about 20°. Introduce the materials for obtaining H into the generator, and attach the washer, as Why should the H apparatus be allowed to stand a short time before igniting the jet? directed in Exp. 16, page 32. Also the jet, as in Exp. 21, page 38. Allow the apparatus to stand four or five minutes,

C

in order to expel the air; then ignite the H as it issues from the jet. Bring the O jet within one eighth of an inch of the burning H, and about the same distance above the point from whence it issues. Place a small piece of board or some other weight upon the gas-bag, and open the stop-cock. The two gases will now burn together. (See Fig. 20.) The apparatus thus arranged constitutes an oxy-hydrogen blow-pipe.

What does the apparatus thus arranged constitute?

Fig. 20.

Exp. 33. Hold in the flame, by means of a pair of pliers, a fine piece of copper wire not more than an inch long. It will burn with a beautiful green flame. O has united with copper.

What colored flame does burning copper produce?

Exp. 34. Instead of the copper, use

How does iron wire burn?

iron wire. It will burn with scintillations. Protoxide and sesquioxide of iron are formed by the combination.

Exp. 35. Treat in the same manner a

How does platinum burn?
Can platinum be melted by the heat of the most powerful furnace?

piece of platinum wire. It will burn with a delicate white light. This metal can not be burned or even melted by the heat of the most powerful furnace, yet it burns readily in the oxy-hydrogen flame.

Exp. 36. Whittle a piece of unslaked lime to a sharp point, and hold it in the

When lime is held in the oxy-hydrogen flame, what is produced?
What is the light called?
Why is it called the Drummond light?

flame. A light will be produced nearly as dazzling as the sun. It is called "the Drummond Light," from its having been discovered by Lieutenant Drummond, of the British navy.

Exp. 37. Fill a bell-glass with the

What takes place when H and O are mixed together, by volume, two of H and one of O, and soap-bubbles filled with the mixture are ignited?

mixed gases, by volume two of H and one of O. Transfer to the gas-bag, attach the bubble-pipe, and fill soap-bubbles with the mixed gases.* By the aid of an assistant, each bubble may be ignited as it ascends. A loud report will follow. The O and the H contained within the bubble are now

---

* Great caution should be observed in performing experiments with the mixed gases. The bubble should not be ignited until it is at least two feet from the pipe, otherwise the fire might be communicated to the gas-bag, when the whole would explode at once.

Fig. 21.

What is the re-sulting com-pound?

chemically united, and the resulting compound is HO (water). (See Fig. 21.)

What, then, do we learn?

Hence we learn that if H is mechanically mixed with O, by volume two of H and one of O, it forms an explosive mixture.

In what propor-tion do these gases burn to-gether?

What compound is formed?

What causes the intense heat?

The two gases, as has already been seen, burn together in the same proportions, and the resulting compound is the same in both cases, HO (water). The cause of the intense heat produced when the two gases burn together may be found in

How much do the gases condense when they unite chemically?

the condensation of the gases as they unite. Whenever bodies condense, heat is set free, and, during the chemical union of H and O, the gases are condensed three thousand times; that is, three thousand pints of the mixed gases form only one pint of water.

How many atoms of H unite with one atom of O to form water? How much by volume? How much by weight?

One atom of H unites with one atom of O to form water; by volume, two of H and one of O; and by weight, one of H and eight of O. Now, if one pint of O weighs eight times more than two pints of H, then one pint of O must weigh *sixteen times* more than *one* pint of H. Therefore O is sixteen times heavier than H.

How much heavier, then, is O than H?

LESSON XII.

What is the equivalent of chlorine? Its symbol?

Chlorine. Equivalent 35. *Symbol Cl.*

Exp. 38. Introduce into a small glass flask two ounces of HCl (hydrochloric acid), and about one third of an ounce of MnO^2 (peroxide of manganese). Adapt a cork to the flask, and also a glass tube bent at right angles; attach another similar tube by means of an India-rubber connector, so as to conduct the element set free downward. Place the flask in the sand-bath, and apply a gentle heat, when Cl will be rapidly disengaged, and may be collected in six-ounce bottles for use. The bottles should have wide mouths and ground stoppers. One end of the tube

Describe the apparatus used in forming it.

What kind of bottles are used for collecting Cl?

should pass to the bottom of the bottle,

Why does the gas expel the air? and the gas, being heavier than air, will soon expel it. (See Fig. 22.) When the

Fig. 22.

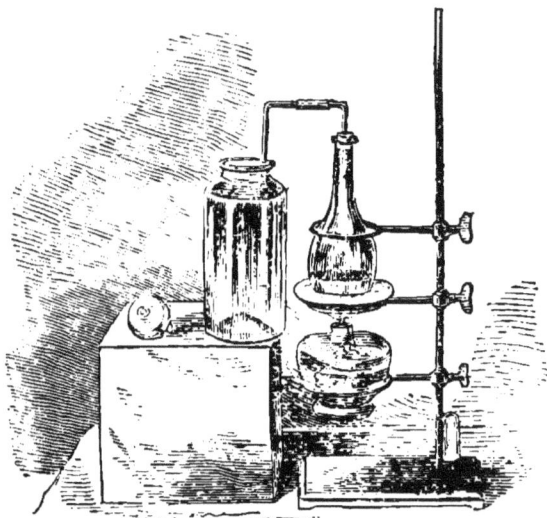

bottle is filled with a light green substance, it may be removed, and carefully

How is the bottle to be closed when filled with it? closed with the ground stopper smeared with tallow. Another bottle may now be filled in a similar manner. A large rag,

Why wave a rag wet with ammonia and alcohol in front of the face while experimenting with Cl? wet with alcohol and liquid ammonia, constantly waved in front of the face, will prevent any injurious effects which might occur from breathing minute portions of this noxious element.

What materials were used? The materials used for obtaining Cl were MnO^2 (peroxide of manganese) and HCl (hydrochloric acid). When MnO^2 is

When hydrochloric acid is brought in contact with peroxide of manganese, what becomes of the O? With what does the Cl unite? The compound formed? How is one atom of Cl obtained from the bichloride of manganese?

brought in contact with HCl, the O of the MnO^2 unites with the H of the HCl, and forms HO (water), while the Mn unites with the Cl, and forms $MnCl^2$ (bichloride of manganese). This $MnCl^2$ readily gives up one atom of its Cl on applying heat, when the bichloride of manganese is reduced to a protochloride. $MnCl^2 = MnCl$; $+ Cl$ set free.

Fig. 23.

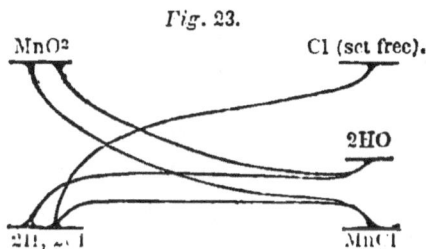

Draw and explain Fig. 23.

Exp. 39. Fill a small-necked bottle one third full of water, and the remaining space with Cl. Wet the thumb, and place it firmly upon the mouth of the bottle, and shake it briskly. The thumb will be

Why is the thumb pressed into the bottle in Exp. 40?

pressed into the bottle, because the water has absorbed the Cl and produced a vac-

How much Cl will water absorb? What is the solution called?

uum. Water absorbs twice its own bulk of Cl. This solution is called *Chlorine water.*

If calico be introduced into Cl, will its colors be affected? Does Cl act upon all animal and vegetable colors?

Exp. 40. Introduce a piece of calico, slightly moistened with water, into a jar of Cl; its color will disappear. Cl bleaches all animal and vegetable colors.

Exp. 41. Drop a few drops of Cl water into some ink, its color will be discharged.

What effect does Cl water have upon ink?

Exp. 42. Perfume a white handkerchief with otto of rose, and drop some Cl water upon it. Fold and press it a few times between the hands, and the odor will be destroyed. Cl destroys all perfumes, whether offensive or otherwise.

How does it act upon perfumes of all kinds?

Malarious matters which communicate disease are rendered harmless by it; and the sick-room is rendered pleasant by sprinkling Cl water over the floor.

How upon malarious matters?

How may a sick-room be rendered pleasant?

Exp. 43. Take as much pulverized metallic antimony as can be held between the thumb and finger. Remove the stopper from a bottle of Cl, and drop in the metal. It will glow as it passes downward through the gas. The bottle will be filled with a white fume, which is chloride of antimony. Cl will unite directly with the metals.

Describe the experiment with antimony.

With what will the bottle be filled? What is the name of the compound? Will Cl unite with the metals?

Exp. 44. Put a small piece of gold-leaf into some Cl water. It will soon disappear. The element Cl has united with the element gold, and formed the compound chloride of gold.

Describe the experiment with gold-leaf.

What is the compound called?

Exp. 45. Wet a rag with oil of turpentine, and immerse it in a jar of Cl. It will burn spontaneously. Cl supports combustion, but not so perfectly as O.

Describe the experiment with oil of turpentine.

Does Cl support combustion?

Exp. 46. Drop a piece of the metal Na (sodium), of the size of a pea, into a gill of strong Cl water. A combination will take place. Evaporate the solution nearly to dryness, and allow it to cool; cubical crystals will appear. The element Cl has united with the element Na and formed NaCl (chloride of sodium), which is common salt. Thus we see that Cl unites directly with the metals to form salts, called *Chlorides.* It is a gas of a light green color, and is about once and a half heavier than air. It has a suffocating odor, and acts as a deadly poison when breathed in any considerable quantity. Consumptive patients, however, are said to have obtained temporary relief by breathing it in very minute portions, mixed with a large quantity of air.

Describe the experiment with sodium.

With what has the Cl united?
What is the chemical name of the compound?
Its common name?
With what does Cl unite to form salts?
What are these salts called?
What are the properties of Cl?

What is said of consumptive patients?

LESSON XIII.

Hypochlorous Acid. Equivalent 43.
Symbol Cl O.

The symbol of hypochlorous acid?
Its equivalent?

Exp. 47. Place a few crystals of KO, ClO⁵ (chlorate of potassa) in a beaker or wine-glass, and cover them with HCl (hydrochloric acid). The upper part of the glass will soon be filled with a gas which

What are the materials used for obtaining it?
In what are they placed?

C 2

What color has
the gas?
resembles Cl (chlorine), though its color
is somewhat brighter. It is ClO. 2 atoms of HCl and 1 atom of KO, ClO^5 are

Give the formula
illustrating the
decomposition
upon the black-
board.
Does the gas ever
explode sponta-
neously?
converted into 2HO ; KO and 3ClO.
$2HCl ; KO, ClO^5 = 2HO ; KO ; 3ClO.$
Sometimes the gas explodes spontaneously. Hence the experimenter would do
well to stand at a little distance while it
is forming.

Exp. 48. Wet a pine shaving with spir-

Explain Exp. 48.
its of turpentine, and immerse it in the
gas by means of a bent wire. It will
burst into spontaneous combustion.

Exp. 49. Fill another glass with ClO,
as in Exp. 48. Attach a small bit of P

If phosphorus be
brought in con-
tact with ClO,
what phenome-
non is presented?
(phosphorus) to the end of a bent wire,
and immerse it in the gas. A spontaneous explosion will follow.

Give the symbol
and equivalent of
chlorous acid.
Chlorous Acid. Equivalent 67. *Sym-*
bol ClO^4.

What materials
are used for ob-
taining chlorous
acid?
Exp. 50. Use SO^3, HO (sulphuric acid),
instead of HCl (hydrochloric acid), as in
Exp. 47. The wine-glass will be filled

What is its color?
in a short time with a gas of a still bright-

Why should the
experimenter use
great caution in
forming chlorous
acid?
er color than ClO (hypochlorous acid). It
is ClO^4, which is very explosive, and
should be used with great caution.

Exp. 51. Immerse a piece of calico,
moistened with a little water, in ClO^4

What is said of calico immersed in this gas? What is the peculiar characteristic of hypochlorous or chlorous acid?

(chlorous acid). It will soon be bleached. Cl (chlorine), ClO (hypochlorous acid), and ClO^4 (chlorous acid), possess powerful bleaching properties.

The symbol and equivalent of chloric acid?

Chloric Acid. Equivalent 75. Symbol ClO^5.

How is chloric acid obtained?

Exp. 52. Dissolve some chlorate of baryta in water, and dilute some sulphuric acid with two parts of water, and add it gradually to the first solution. A white powder will be thrown down, which is sulphate of baryta. The ClO^5 is set free, and

Of what is chlorate of baryta composed?

is held in solution. Chlorate of baryta is composed of chloric acid and baryta. Sul-

Which exhibits the stronger affinity for the baryta, the sulphuric or the chloric acid?

phuric acid has a stronger affinity for the baryta than has the chloric acid, and, by single elective affinity, the sulphuric acid unites with the baryta, while the chlo-

Which acid is liberated?

ric acid is liberated. Sulphuric acid $= SO^3, HO$. Chlorate of baryta $= BaO, ClO^5$.

Give the formula upon the blackboard.

BaO, ClO^5; $SO^3, HO = BaO, SO^3$; HO;

If chloric acid, in this dilute state, is allowed to evaporate spontaneously, what will it eventually become?

ClO^5. ClO^5 (chloric acid), in this dilute state, if allowed to evaporate spontaneously, will eventually become a yellowish oily liquid, when it is said to be concentrated.

If a slip of newspaper be dipped into concentrated chloric acid, what follows?

Exp. 53. Dip a slip of newspaper into concentrated ClO^5. In a short time it will burn. ClO^5 parts with some of its

What causes the combustion? / Does this acid possess bleaching properties? O, which combines with the combustible matter. This acid does not possess bleaching properties like the preceding compounds of Cl.

What is the symbol of perchloric acid? Its equivalent? *Per-* or *Hyperchloric Acid. Equivalent* 91. *Symbol* ClO^7. ·

What does this acid resemble? This acid resembles the preceding in properties, with the exception that its affinities are much stronger. None of the

Are the compounds of Cl well understood? compounds of Cl are at present well understood. An intermediate acid, called

What is said of hypochloric acid? hypochloric acid, ClO^3, partially described by Millon, has been omitted on account

Has its existence been well established? of its existence not having been well established.

LESSON XIV.

Give the equivalent and symbol of nitrogen *Nitrogen (Azote). Equivalent* 14. *Symbol N.*

How is it obtained? *Exp.* 54. Place a capsule (Fig. 17) upon a soup plate filled with water. Dry a small piece of P (phosphorus) carefully between blotting-paper, and lay it upon

Describe Exp. 54. the capsule. Ignite it, and quickly invert over it a half-gallon candy-jar, which will soon be filled with dense white fumes. P has united with the O of the air contained in the jar, and formed PO^5 (phosphoric acid), while the N (nitrogen) of the air is left free. The air contained in the jar

Was the air contained in the jar a chemical compound?
was not a chemical compound, but a mechanical mixture. Hence the following diagram will only show the separation of

Fig. 24.

Explain the difference between a chemical compound and a mechanical mixture.

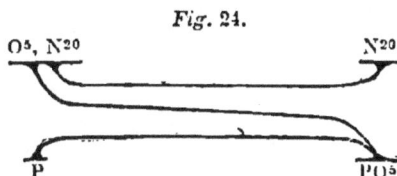

What becomes of the white fume in the jar?
a mechanical mixture. The white fume will soon be absorbed by the water in the

What portion of the jar will be filled with water, and what with gas?
plate, when about four fifths of the jar will be filled with a colorless gas, the other fifth being occupied with water.

Exp. 55. Place the jar containing the N in an upright position upon the table, and lower into it a lighted candle. The

Does nitrogen support combustion?
combustion will cease. N does not support combustion.

Exp. 56. Fill a smaller jar with N by pouring the gas upward into it under water. The receiving jar should first be filled with water, and raised with the mouth

How may gases be transferred from one jar to another?
downward until its mouth is near the surface. The jar containing the N may now be depressed in the water, and its mouth so directed that the gas will escape into the smaller jar. (See Fig. 25.)

Fig. 25.

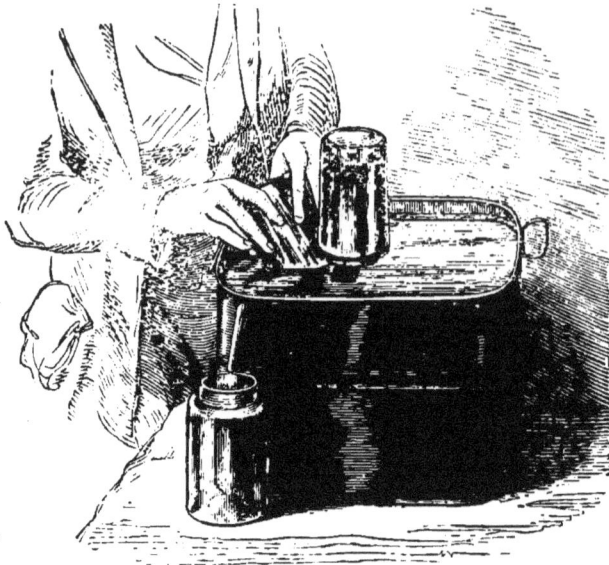

Does nitrogen support respiration?

How was this proved?

Does the animal die from the poisonous effects of the gas?

What name was improperly applied to this element?

How much nitrogen do we constantly breathe?

Exp. 57. Drop a living mouse into the jar filled with N. He will soon die. N does not support respiration. The animal does not die, however, from any poisonous effects of the gas, but from a want of O. Hence the name *azote* (life destroyer), by which this element was formerly called, was improperly applied. We are constantly breathing about four times as much N as O, without witnessing any poisonous effects from it. N is negative or passive in its effect upon the animal economy. All animals die when deprived of O; and in crowded assemblies and unventilated sleeping apartments, the air

What observation is made of crowded assemblies and unventilated sleeping apartments!

Should air be breathed a second time?

gradually becomes poor in O from the quantity of it consumed in breathing. Air should not be breathed a second time. We shall hereafter see that the O, when exhaled, has entered into a combination in which it acts as a deadly poison when brought in contact with the lungs. Hence

What, then, do we learn of nitrogen!

we learn that N is a colorless gas having negative properties. It is neither combustible, a supporter of combustion, nor a supporter of animal life. It constitutes about four fifths of the bulk of the atmosphere.

LESSON XV.

What is the equivalent of protoxide of nitrogen? Its symbol!

Protoxide of Nitrogen, or Exhilarating Gas. Equivalent 22. *Symbol* NO.

Exp. 58. Introduce two ounces of NH^3, NO^5 (nitrate of ammonia) into a pint glass

How is it obtained!

retort, arranged as in the formation of O, page 42, Fig. 15. Apply heat by means of the spirit-lamp and sand-bath. First

What first takes place with the salt! What follows?

the salt will melt, ebullition will follow, and when the temperature is sufficiently high, a rapid decomposition will take

Of what is the salt composed!

place. The salt is composed of nitric acid and ammonia.

Nitric acid $= NO^5$. Ammonia $= NH^3$.

Fig. 26.

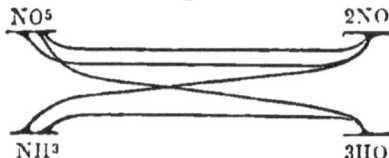

Explain the decomposition by means of the diagram, Fig. 26.

Three atoms of O from the NO^5 united with the three atoms of H of the NH^3, and formed 3HO (water), while the two atoms of N united with the two remaining atoms of O, and formed 2NO (protoxide of nitrogen).

Exp. 59. Repeat Exp. 28 in NO instead of O. The candle will be relighted, but will not burn so brilliantly as in O. NO supports combustion more vividly than air.

If a lighted candle be immersed in protoxide of nitrogen, will the flame be increased?
Does NO support combustion more vividly than air?

Exp. 60. Place a mouse in a jar of NO. He will exhibit signs of pleasure. NO supports animal life.

Does it support animal life?
Explain Exp. 60.

Exp. 61. Fill a gas-bag, containing at least one gallon, and furnished with a mouth-piece and stop-cock, with NO, by means of the bell glass and pneumatic trough.* (See Fig. 19, page 49.) Permit some student of an active temperament to inhale the gas, which should be done in

Explain the manner of inhaling NO.

* The gas should be allowed to stand over water at least two hours before it is used, in order to absorb its impurities.

the following manner: The student should exhaust his lungs of all air, hold his nose, and fill his lungs from the gas-bag. The gas should be breathed out and in four or five times, in order to obtain its happiest effects. A glow of excitement will overspread the whole system. Some laugh immoderately, others play the orator, while others pass through with some devotional exercise. A few subjects have been found, who, when under the influence of NO, exhibited pugilistic propensities, particularly when the gas was impure, or when the subject's mind had been previously excited. The opinion, however, that the gas invariably develops the leading traits in the character of the person who inhales it is incorrect, as the subject never entirely loses consciousness when properly under its influence. Sir Humphrey Davy breathed it for five minutes from a gas-bag containing nine gallons without losing his consciousness. All persons, except those who constantly stimulate with opium or spirituous liquors, are powerfully excited by breathing NO for the space of about thirty seconds. Persons of plethoric habit should not inhale the gas, as it excites an increased circulation of the blood. NO (protoxide of ni-

What effects are produced upon the system?

Does the gas always develop the leading traits of character of the person who inhales it?

Does the subject entirely lose consciousness?

Give the experiment of Sir H. Davy.

What class of persons are not affected by breathing NO?

Why should not persons of plethoric habit inhale this gas?

trogen), as we have seen, is a colorless gas. It has a sweet taste, supports com-
What properties of NO have been developed by the foregoing experiments?
bustion and animal life, but will not burn itself. When animals breathe it, they live faster than when breathing common air, as they obtain more of the life principle, O.

Give the equivalent and symbol of binoxide of nitrogen.

Binoxide of Nitrogen, or Nitric Oxide.
Equivalent 30. *Symbol* NO^2.

Exp. 62. Place some copper slips, or a cent, in a bottle containing four or five ounces, to which has been adapted a cork and bent tube. Pour in an ounce of NO^5,
How is this compound obtained?
HO (nitric acid) and one fourth the quantity of HO (water). NO^2 will be set free with effervescence, which may be collected in a small glass jar over water. Slide the jar, when filled, upon a plate filled with water, and transfer to the table for use.

Explain Fig. 27.

Fig. 27.

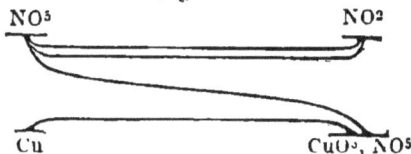

One equivalent of the NO^5 (nitric acid) is decomposed. An atom of Cu unites with three atoms of its O, forming CuO^3 (teroxide of copper), while one atom of NO^2 (binoxide of nitrogen) is set free. An undecomposed atom of NO^5 now combines with the CuO^3 thus formed, and the action is repeated until some of the materials are exhausted.

Exp. 63. Gradually admit some air into a jar of NO^2. Dense reddish fumes

will soon fill the jar. The NO^2 has elected two more atoms of O from the admitted air, and formed NO^4 (nitrous acid). NO^3 (hyponitrous acid), at common tem-

peratures, is a bright green volatile liquid, but its properties are little known. All the chemical compounds of O and N (with the exception of NO) are deadly poisons when breathed.

LESSON XVI.

Nitric Acid.   *Equivalent* 63.   *Symbol* NO^5, HO.

The symbol of this compound, if it could be obtained in the dry state, would

be NO^5, and its equivalent would be 54. It has never yet been obtained in this state, but it is probable it thus exists, and will hereafter be thus obtained. Dry or anhydrous nitrate of potassa or ammonia is composed of NO^5, KO, or NO^5NH^3. But the NO^5 has not yet been separated from either of these bases without the presence of water, one equivalent of which it always takes with it, when its formula is NO^5, HO.

Exp. 64. Introduce into a half-pint tu-

How is nitric acid obtained?

bulated glass retort one ounce of NO^5, KO (nitrate of potassa or saltpetre), and two ounces of SO^3HO (sulphuric acid). The beak of the retort should be adapted to a half-pint receiver by means of a perforated

Describe the apparatus.

cork. The upper mouth of the receiver should be closed by a cork, through which a small glass tube should pass quite to the bottom. This tube is called a *safety*

What is a safety-tube?

tube. Surround the receiver with cold water, and apply a gentle heat to the re-

When heat is applied, what is decomposed?

tort by means of the sand-bath and spirit-lamp. Decomposition of the KO, NO^5 (nitrate of potassa) will take place. KO,

Give the formula of the substances used before they are applied. Give it afterward.

NO^5 ; $2SO^3HO$ are resolved into $2SO^3$, KO ; NO^5, HO ; HO. The stronger, SO^3,

Fig. 28.

Draw and explain the diagram.

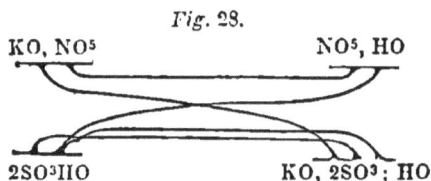

KO, NO^5 NO^5, HO

$2SO^3HO$ KO, $2SO^3$; HO

As the NO^5 is set free, what body does it take with it?

elects the KO, and the NO^5 is set free, together with an equivalent of HO (water).

What remains in the retort?

One equivalent or atom of HO remains with the KO, $2SO^3$ in the retort, though it is not chemically united. NO^5, HO (nitric acid) must be kept in ground-stopper bottles.

Exp. 65. Add to half an ounce of water three drops of NO^5, HO, into which solution dip a piece of litmus paper.* It will be changed from blue to red. Well-marked acids always produce this result.

Explain Exp. 65.

What effect do well-marked acids always produce upon litmus paper?

Exp. 66. Place a few leaden shot in a wine-glass, and cover them with water. Add twice the quantity of NO^5HO (nitric acid). A reddish fume of a suffocating odor will be disengaged, which is NO^4 (nitrous acid). One atom of O from the NO^5 has united with the Pb (lead), and formed PbO (protoxide of lead). An undecomposed equivalent of NO^5 united with the oxide thus formed and dissolved it, forming PbO, NO^5 (nitrate of protoxide of lead). This latter compound is not soluble in the acid, but is soluble in water.

If nitric acid be brought in contact with lead, what follows? Is the lead (Pb) dissolved?

How much O from the acid first combines with the lead? What is the compound called? With what does an undecomposed equivalent of nitric acid now unite?

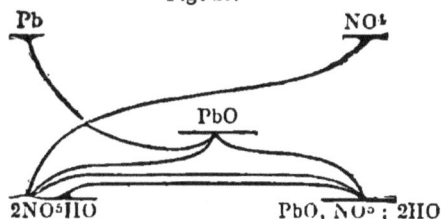

Fig. 29.

Explain the figure.

Hence, if no water were present, it would

* For the preparation of litmus paper, see p. 136.

Why was water added?

remain as a solid incrustation upon the surface of the metal, and thus put an end to the action. See formation of H (hydrogen), page 32.

Exp. 67. Pulverize some charcoal, and dry it thoroughly. Place a spoonful of

If nitric acid be dropped upon pulverized charcoal, what phenomenon is presented?

it in a wine-glass. Tie a small test-tube to a rod six feet long, and pour into it a drachm or two of strong NO^5, HO (nitric

Why should the experimenter stand at a distance when bringing the two bodies in contact?

acid). Stand at a distance, and drop the acid upon the charcoal. A combustion will follow, with the disengagement of dense poisonous fumes, which are chiefly NO^4 and CO (nitrous acid and carbonic oxide).

What results from the action of nitric acid on spirits of turpentine?

Exp. 68. Instead of the charcoal, use a few drops of spirits of turpentine. Another spontaneous combustion will take place.

Exp. 69. To a saturated solution of KO, CO^2 (carbonate of potassa) add

Explain Exp. 69.

NO^5, HO (nitric acid) until effervescence ceases. Place the solution in the sun, or

What is the formula for the salt formed, and by what names is it sometimes called?

some warm place, until prismatic crystals appear. They are KO, NO^5 (nitrate of potassa, nitre, saltpetre).

Fig. 30.

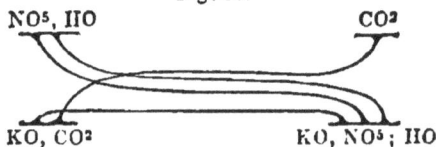

NO⁵, HO   CO²

KO, CO²   KO, NO⁵; HO

Exp. 70. In a Wedgwood mortar place
six parts by weight of KO, NO⁵ (nitrate
of potassa), one part of S (sulphur), and
one of C (carbon, charcoal) : pulverize the
whole thoroughly for ten minutes. Now
add water sufficient to form a paste, and
mix thoroughly. Perforate a piece of tin
or lead, and press the paste through. Allow the threads to fall upon some paper,
and when partly dry, rub the mass gently
with the fingers, and small grains will be
formed, which are *gunpowder.*

Exp. 71. Place some gunpowder upon
a capsule, and ignite it by means of a red-
hot wire or glowing coal. The solid powder is converted into gases, with the exception of a small residue. The gases are
CO² (carbonic acid), and N (nitrogen),
which now occupy a much greater space
than when in the solid form. Hence the
explosion. Formula :

$$3C \, ; \; S \, ; \; KO, NO^5 = KS \, ; \; 3CO^2 \, ; \; N.$$

Fig. 31.

What are the materials used for forming gunpowder?

Describe the process.

When powder is ignited, into what is the solid body converted?

What names are applied to these gases?
Give their symbols and equivalents.
What causes the explosion?
Give the formula before and after the explosion.

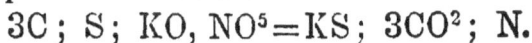

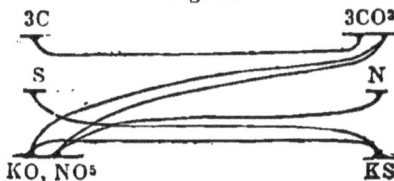

Draw and explain the figure.

LESSON XVII.

The equivalent and symbol of ammonia?

Ammonia. Equivalent 17. Symbol NH^3.

Exp. 72. Perforate a cork with a round file, and insert a small glass tube bent twice at right angles. Fit the cork to a large test-tube, into which introduce one

Describe the experiment for obtaining ammonia.

drachm of NH^3, HCl (sal ammoniac), together with an equal quantity of CaO, HO (slaked lime). Pass the conducting tube to the bottom of a small phial containing half a drachm of HO (water). Apply heat to the test-tube, and NH^3 will be set free. The water in the phial, if

How much of the gas will be absorbed by cold water?
What is the solution called?

kept cold, will absorb near seven hundred times its own bulk of the gas. The solution is called *Aqua Ammonia*, and its common name is *Spirits of Hartshorn.*

Give the formula before and after the decomposition.
Is ammonia an acid?

Formula: NH^3, HCl; CaO, HO = CaCl; 2HO; NH^3 set free in its gaseous form.

Exp. 73. Drop some liquid (*aqua*) ammonia upon red test-paper.* It will be

Is it an alkali?
How was this ascertained?

changed to blue. All alkalies produce this effect; hence NH^3 is an alkali.

* For making red test-paper, see p. 136.

How are harts-horn bottles commonly prepared? Equal parts of slaked lime and sal ammoniac form the compound contained in the *hartshorn bottles* of the shops.

What is the equivalent of carbon? Its symbol? *Carbon. Equivalent 6. Symbol C.*

Exp. 74. Fold a piece of filtering-paper so as to fit it into a funnel. Fill it two thirds full of finely-pulverized C (charcoal). Discolor some water with ink, *How does charcoal affect various coloring matters? Why do sugar refiners use it?* and filter it through the C. It will be rendered colorless. C (charcoal) absorbs various coloring matters, which property renders it useful to sugar refiners.

Exp. 75. Pour half a bushel of charcoal into a cistern containing nauseous *If charcoal be poured into a cistern of nauseous water, what takes place?* water. Within two or three days the water will be deprived of its disagreeable odor. The charcoal has absorbed the un-*To what other purpose has charcoal been applied?* pleasant gases. A cellar containing decaying vegetable matter may be rendered *Is charcoal one of the forms of carbon?* pleasant by the presence of charcoal. C *When carbon is crystallized, what is it called?* (carbon), when crystallized, forms the *diamond.* Charcoal is nearly pure C in a *What of soot and coke?* state of minute division. Soot, or lamp-*Black-lead or plumbago?* black, and coke, are also nearly pure C. Plumbago, or black lead, is another form of C, containing a slight quantity of iron, *For what is this substance extensively used?* and is extensively used for reducing friction in machinery, and in the manufacture of drawing pencils. Crucibles are

D

Why are crucibles made of plumbago?

sometimes made of plumbago on account of its property of withstanding an intense

When exposed to air or water, will charcoal decay?

heat.   Charcoal will not decay when exposed to air or water.   Hence posts which have been charred never decay.   It is

What is said of the stakes driven into the bed of the Thames by the ancient Britons?

said that the stakes* driven into the bed of the River Thames by the Britons in the year 55 B.C., to prevent its passage by Julius Cæsar, when discovered a few years since, were in a perfect state of preservation from having been charred.   Meat

How may meat be preserved fresh in warm weather?

packed in charcoal dust may be preserved fresh for several weeks in warm weather.

Give the equivalent of carbonic oxide. Its symbol?

Carbonic Oxide.   Equivalent 14. *Symbol C O.*

This compound is always formed when wood is burned with an incomplete supply of air.   When charcoal has burned for

How is this gas formed?

some time in air, and ashes have accumulated upon the surface so as to prevent a complete supply of O, CO (carbonic oxide) is invariably formed.   The same result takes place when the damper of a stove is

Does CO possess poisonous properties?

closed.   CO is a gas which possesses very poisonous properties, and when breathed

What sensation is produced on the system when this gas is breathed much diluted with air?

much diluted with air, produces drowsiness, giddiness, and sometimes fainting. The sleepy sensation experienced when

* Grey.

How is the drowsy sensation produced when sitting by a hot stove in winter?

Should rooms ever be heated with a chafing-dish of charcoal?

What two poisonous gases are set free when charcoal burns in air? Which first?

sitting by a warm stove in winter, is usually produced by breathing small portions of CO mixed with air. Rooms should never be heated by a chafing-dish of charcoal, as two poisonous gases are formed by this combustion. First, before the accumulation of ashes, CO_2 (carbonic acid) is formed, after which CO (carbonic oxide). Many valuable lives have been sacrificed upon the altar of ignorance connected with these two noxious compounds.

When persons lodge in rooms heated by burning charcoal, what is the first sensation produced?

If fresh air is not admitted, what ensues?

Persons, when attempting to lodge in rooms warmed by burning charcoal, feel an indescribable inclination to sleep, into which state they soon pass; and if not speedily relieved by the admission of fresh air, death ensues.

LESSON XVIII.

Give the equivalent of carbonic acid?
Also its symbol?

Carbonic Acid. Equivalent 22. Symbol CO_2.

How is this compound obtained?

Exp. 76. Introduce an ounce of NaO, $2CO_2$ (bicarbonate of soda) into the hydrogen generator, and add two ounces of water. Pour through the funnel-tube a small portion of SO_3, HO (sulphuric acid). CO_2 will be set free with effervescence, and may be collected as in Exp. 16, page

Give the formula before and after the decomposition.

32. NaO, 2CO^2 ; SO^3, HO = NaO, SO^3 ; HO ; 2CO^2 set free. The stronger, SO^3, elects the NaO, and drives off the CO^2 in its gaseous form.

Fig. 32.

NaO, 2CO^2 2CO^2

Draw and explain the diagram.

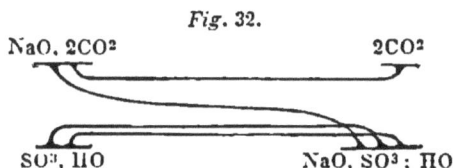

SO^3, HO   NaO, SO^3 ; HO

Exp. 77. Place a jar of CO^2 (carbonic acid) upon the table, with its mouth upward, and lower into it a lighted candle.

Will carbonic acid support combustion? The flame will be extinguished. CO^2 will not support combustion, nor will it Will it burn itself? burn itself.

Explain Exp. 78. *Exp.* 78. Pour the CO^2 from the jar upon a lighted candle. The combustion What does it show? will cease. CO^2 is heavier than air.

Exp. 79. Drop a living mouse into a jar of CO^2. Life will apparently cease. Does carbonic acid support animal life? Does oxygen? What experiment proves this? Remove him quickly, and place him in a jar of O. His life will be restored. CO^2 (carbonic acid) destroys animal life, and O (oxygen) sustains it.

Exp. 80. Place a spoonful of lime in a pint of rain water, and allow it to stand for twenty-four hours, during which time agitate the liquid several times. Pour off the clean solution carefully into a ground-stopper bottle, and keep it for fu-

How would you show that carbonic acid is produced by respiration? ture use. Breathe into a wine-glass containing lime-water, by means of a glass tube, for some time. It will gradually assume a milky appearance. Insoluble CaO, CO^2 (carbonate of lime) is formed, which shows that CO^2 (carbonic acid) is produced by breathing—respiration.

What is the symbol of carburetted hydrogen? Its equivalent? *Carburetted Hydrogen. Equivalent* 28. *Symbol* C^4H^4.

Exp. 81. Adapt a pipe-stem to a cork which fits a large test-tube. Place half How may this gas be easily obtained? an ounce or less of mineral coal in the test-tube, press in the cork, and apply heat. C^4H^4 or CH (carburetted hydrogen) will **soon** be forced through the pipe- What is the color of its flame? Is it city gas? stem. Ignite it. It burns with a white light. This is common city gas, mixed with impurities.

Exp. 82. Introduce one and a half fluid ounces of alcohol, and three of strong SO^3, HO (sulphuric acid) into a pint re- Explain Exp. 82. tort, and apply a gentle heat. At first the solution assumes a black appearance, but when the temperature is sufficiently high, a rapid decomposition follows, with What gas is disengaged? Of what is alcohol composed? Sulphuric acid? Give the formula illustrative of the decomposition. the disengagement of C^4H^4 (carburetted hydrogen). Alcohol is composed of $C^4H^6O^2$. Sulphuric acid of SO^3, HO. $2SO^3HO$; $C^4H^6O^2 = 2SO^3$, HO; 2HO;

How may the gas be collected? C^4H^4. The gas may be collected over water, which absorbs only about one eighth of its own volume of it.

Exp. 83. Fill a bell glass with C^4H^4 (carburetted hydrogen), and transfer to the gas-bag. Attach the jet, and by means of a weight, force out the gas,

Of what color is the flame produced by burning carburetted hydrogen? which may be ignited. It burns with a beautiful white light. It is the same gas that was formed by Exp. 81, but contains less impurities, from having been passed

What is the result of the combustion? through water. The result of the combustion of C^4H^4 in air is CO^2 (carbonic acid) and HO (water). C^4H^4 has been

Why was this compound called olefiant gas? called olefiant gas, because when mixed with its own volume of Cl (chlorine) the two gases disappear, and an oily liquid is formed.

Fig. 33.

The diagram, Fig. 33.

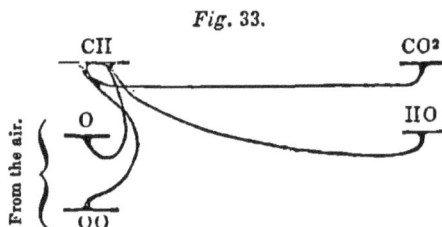

What is the equivalent of light carburetted hydrogen? Its symbol?

Light Carburetted Hydrogen. Equivalent 8. Symbol CH^2.

When is this gas generated spontaneously?

This gas is generated spontaneously in many coal-beds, and when mixed with

When mixed with air, what is it called?

air it is commonly called *Fire-damp*,

Why was it a terror to the miners?

Who has done much to obviate the danger?
What is the construction of Sir H. Davy's safety-lamp.

When vegetable matter decays under water, what gases are formed?

If the muddy bottom of a shallow pond be disturbed, what body will arise to the surface of the water?
Are these bubbles combustible?

When they burn, what is the result of the combustion?

which was a terror to the miners, on account of its explosive properties, until Sir Humphrey Davy invented the safety-lamp, which is simply an oil-lamp surrounded by wire-gauze, which prevents the flame from contact with the gas. CH_2 and a small quantity of CO_2 are always formed when vegetable matter decays under water.

Exp. 84. Disturb the muddy bottom of a shallow pond in summer. Bubbles of gas will arise to the surface, which may be collected in jars, or ignited upon the surface of the water, when they will burn with a yellow light. The result of the combustion is CO_2 (carbonic acid) and HO (water). CH_2; O_4 from the air $= CO_2$; 2HO.

Give the equivalent and symbol of cyanogen.

Is it an element or a compound?
Does it play the part of an element?
What is the most convenient symbol of cyanogen?
What is the general rule?
Does this compound ever unite chemically with single elements?

Cyanogen. Equivalent 26. *Symbol* C_2N *or Cy.*

This compound of C and N plays the part of an element, and its most convenient symbol is Cy. As a general rule, elements only unite with elements, and compound bodies with compounds. Cy is an exception to this, as it combines with several of the elements separately.

Hydrocyanic Acid (*Prussic Acid*).
Equivalent 27.   Symbol CyH.

The compound body Cy (cyanogen) unites with the simple body H (hydrogen), and forms the most vigorous of all the poisons, one drop of the liquid upon the tongue of a dog being sufficient to produce death in a few seconds. This and the preceding compound should be experimented with only by experienced chemists.

LESSON XIX.

Sulphur.   Equivalent 16.   Symbol S.

Exp. 85. Fill a test-tube half full of flowers of S (sulphur). Heat it until it melts. It is now liquid. Pour one half of it into cold water. It is now solid. Heat the remainder still more strongly, and it will be converted into vapor. Conduct this vapor into a cold vessel, and it will be condensed into *Flowers of S.* Bodies exist in one of three states, the liquid, the solid, or the gaseous. S, by this experiment, is made to assume all three.

Exp. 86. Place an ounce and a half of

iron filings and an ounce of flowers of S in a crucible, which should be supplied with a cover, and heated to redness. The S will soon unite with the Fe (iron), forming FeS (sulphuret of iron). This compound is found in great abundance in nature, and is sometimes called *iron pyrites*, at others *fool's gold*. It has, when found native, a lustre very much resembling this metal.

How may sulphuret of iron be formed artificially?

Is this compound ever found free in nature?

What names have been applied to it?

What metal does it resemble?

Give the symbol of sulphuretted hydrogen. Its equivalent.

Sulphuretted Hydrogen (Hydrosulphuric Acid). Equivalent 17. Symbol HS.

Exp. 87. Introduce half an ounce of FeS (sulphuret of iron) into a chemical flask holding half a pint, to which has been adapted a cork and bent tube; add half an ounce of HO (water) and an ounce of SO³, HO (sulphuric acid). Apply a gentle heat, and a gas having the odor of rotten eggs will be disengaged. It is HS (sulphuretted hydrogen).

How may this compound be obtained?

What is its odor? Is it a gas, liquid, or solid?

Exp. 88. Conduct HS as it is set free into a ground-stopper bottle containing two ounces, which should be filled half full of HO (water). The tube should scarcely pass beneath the surface. When the water has absorbed three times its own bulk of HS, it will be saturated; that

How much of the gas will water absorb?

D 2

What is the solution called? is, it will receive no more. The solution is called *Sulphuretted Hydrogen Water.*

Exp. 89. Place a bright five cent piece, a piece of lead, and a piece of iron, upon the table. Drop a single drop of the solution of sulphuretted hydrogen upon each.

Illustrate Exp. 89. A black AgS (sulphuret of silver) will be formed upon the coin, a black PbS (sulphuret of lead) upon the lead, but the iron will remain untarnished. In the case of

Give the formulas upon the black-board. the Ag (silver), HS; Ag$=$AgS; H set free. In the case of the Pb (lead), HS; Pb$=$PbS; H set free. Or, by diagram, thus:

Fig. 34.

II S   II (set free).

Ag   Ag S

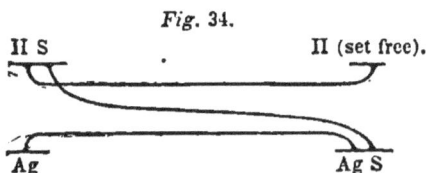

Also the diagrams.

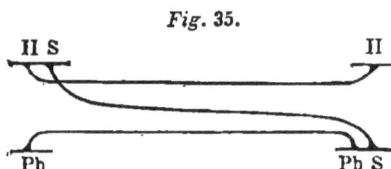

Fig. 35.

II S   II

Pb   Pb S

These experiments show that S has a stronger affinity for silver and lead than Has sulphur an affinity for nearly all the metals? for iron. It has more or less affinity for nearly all the metals.

Exp. 90. Treat solutions of acetate of Explain Exp. 90. lead and nitrate of silver with a solution of HS (sulphuretted hydrogen). A black

What is formed? sulphuret of the metal will be formed in both cases.

What is the equivalent and symbol of hyposulphurous acid?

Hyposulphurous Acid. *Equivalent* 24.
Symbol S O.

Does this compound exist in a separate state?

This compound is not known in the separate state, but its existence is estab-

With what is it always found in contact?

lished. It is always found in contact with a *salifiable base*, as NaO (soda).

Give the equivalent and symbol of sulphurous acid.

Sulphurous Acid. *Equivalent* 32.
Symbol SO^2.

Exp. 91. The apparatus for obtaining NO^5, HO (nitric acid), page 68, will be found quite convenient for this experiment. Place in the retort three fourths

What materials are used for obtaining this compound?

of an ounce of Cu (copper filings), together with an ounce of SO^3, HO (sulphuric acid). The beak of the retort should pass into the receiver until it comes in contact with the safety-tube, which may now be removed, and water poured into the receiver until the beak of the retort is just covered.* Insert the cork and tube, and apply a gentle heat to the retort. SO^2

What is the form of sulphurous acid?

(sulphurous acid) will be set free in its gaseous form, and will be absorbed by the

---

* Instead of the glass stopper, it is safer to close the retort with a cork, through which a small glass tube is made to pass to the bottom.

How much of the gas will water absorb? Water in the receiver until it has received nearly fifty times its own bulk of the gas. As soon as the water commences to rise in the safety-tube the receiver should be disconnected from the retort, and the heat What effect does the gas have upon those who breathe it? removed also, or the gas will be set free in the room, which invariably produces violent coughing. The Cu (copper) elects one atom of O from the SO^3, forming CuO (protoxide of copper), and liberating an Illustrate the decomposition. atom of SO^2, when an undecomposed atom of SO^3 unites with the oxide thus formed, giving the compound CuO, SO^3 (sulphate of protoxide of copper — blue vitriol). Another clean metallic surface is now presented to the SO^3, and the ac- Why is the HO omitted in the diagram? tion is repeated. Omitting the HO of the SO^3 (which is inert in the decomposition), the diagram will read thus:

Draw the diagram and explain it.

Fig. 36.

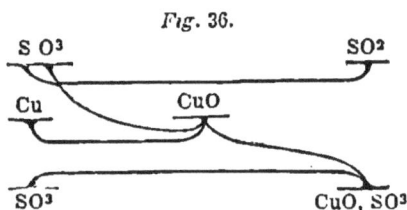

The solution of SO^2 (sulphurous acid) must be transferred to a ground-stoppered bottle, and carefully closed till wanted for use.

Exp. 92. Pour some solution of SO^2

(sulphurous acid) upon a peony flower. Its color will be discharged. SO^2 destroys vegetable colors.

Exp. 93. Place a teaspoonful of S (sulphur) upon a capsule. Ignite the S, and

SO^2 will be formed. Hold over the flame any beautifully colored flower, slightly moistened with water. It will soon be bleached.

Exp. 94. Hold a lighted shaving over

burning S. The flame will cease. SO^2 does not support combustion.

Exp. 95. Hold some small animal over burning S. It will soon die. SO^2

does not support animal life (respiration). Bleachers of straw use large quantities of SO^2.

LESSON XX.

Sulphuric Acid. Equivalent 49. *Symbol* SO^3, HO.

THE equivalent of sulphuric acid, as described by authors in general, is 40. But it is evidently incorrect, as SO^3 is without acid properties until it is combined with an equivalent of HO, or, at

least, until H is present. If SO^3 be sulphuric acid, then 40 must be its equivalent. But if SO^3, HO is sulphuric acid, 49 is its equivalent number. The student will bear in mind, however, that the latter is the true formula of the acid.

Exp. 96. Place some anhydrous F^2O^3, SO^3 (sulphate of sesquioxide of iron) in a large test-tube, furnished with a conducting tube and cork smeared with sweet oil. Apply a strong and continuous heat to the test-tube, and conduct the vapor set free into a cold dry receiver, allowing the first portions to escape into the air. The cold receiver will condense the vapor into a white crystalline solid. It is SO^3, which authors call sulphuric acid. Teroxide of sulphur, if adopted generally, would be a name which would leave no doubt upon the mind of the student as to the proper formula of the compound.

Describe the experiment for obtaining SO³.

If teroxide of sulphur were adopted for SO³, what would be the result?

If SO³ be dropped into water, what takes place?

Exp. 97. Take some SO^3 (teroxide of sulphur) upon a glass rod, and drop it into some water. It will combine vigorously with an atom of the water, causing a hissing sound. SO^3 united to HO, atom to atom, forms the common *oil of vitriol*, or *sulphuric acid* of commerce.

When SO³ unites with HO, what compound is formed?

Exp. 98. Fill the cup of a copper deflagrating spoon (Fig. 37) with S (sulphur). Place half an ounce of HO (water) in a half-gallon candy-jar. Ignite the

S, and allow it to burn in the jar, which should be kept covered. The Fig. 37.
combustion will soon cease, and the jar will be filled with a white fume, which is SO^2 (sulphurous acid). Remove the S dexterously, and quickly

Fig. 37.

Explain Exp. 98 in full.

re-cover the jar, so as to exclude the air. Tie a slip of canton flannel to a glass rod, and saturate it with NO^5, HO (nitric acid). After which, lower it into the fume in the jar. A reddish vapor, of a suffocating odor, will be disengaged, which is NO^4 (nitrous acid). The SO^2 has elected an atom of

Give the formula illustrating the decomposition.

O from the NO^5, and is now SO^3. SO^2; $NO^5 = SO^3$; NO^4. The water in the jar will soon combine with the SO^3, when the burning S may be again introduced, and the whole process repeated several times. In this way some diluted SO^3, HO may be obtained. This compound

How is sulphuric acid prepared for commercial purposes?

is prepared for commercial purposes by heating S in a furnace, and conducting the SO^2 (sulphurous acid) thus formed through the vapor of NO^5, HO (nitric

What does the nitric acid yield?

acid). The NO^5 yields up one atom of

Why is the compound then brought in contact with steam?

its O to the SO^2, forming SO^3, which is then brought in contact with steam, from

What then becomes of the SO^3, HO?

which it takes an atom of water. This latter compound, SO^3, HO, is now absorbed by water upon the leaden floor of the chamber in which the last process is con-

What is afterward done with the watery solution? ducted. The watery solution is afterward evaporated in large glass or platinum retorts, until all the HO (water) has passed off except an atom to each atom of SO^3.

What is the common name of SO^3, HO? It is now SO^3, HO, the common *Oil of Vitriol.*

What takes place when wood is brought in contact with sulphuric acid? *Exp.* 99. Immerse the end of a shaving in SO^3, HO (sulphuric acid). It will soon be charred. The shaving is com-

Why is it charred? posed of C (carbon), O (oxygen), and H (hydrogen), CHO. The acid combines with the HO, and the C is left in a mi-

For what is this process often resorted to by farmers? nute state of division. This process is often resorted to by farmers for charring stakes and posts which are to be used in fencing.

Exp. 100. Incline a test-tube, and slide in a shingle nail, over which pour some water. Add a few drops of SO^3, HO.

Is it known whether sulphuric acid is composed of SO^3, HO, or SO^4, H? What is known in relation to this point? It may be observed here that it is not known whether sulphuric acid is composed of SO^3, HO, or SO^4, H. But it is known that the acid consists of one atom of S, four of O, and one of H. In this

In Exp. 100, what does the iron displace? Into what is SO^4, H, or SO^3, HO converted? experiment the metal Fe (iron) displaces the H, which passes off in bubbles. SO^4, H, or SO^3, HO and Fe is converted into SO^4, Fe, or FeO, SO^3. In either case, it will be seen that the metal Fe has simply taken the place of the H of the acid, while the salt, *sulphate of oxide of iron,* was formed. As soon as bubbles cease to rise from the liquid, remove the nail, and boil the liquid until a drop of it upon a slip of cold glass gradually assumes the crystalline form. It is crystallized sulphate of oxide of iron. Its common name is *Copperas.*

What does the iron displace?

What salt was formed?

What is its common name?

LESSON XXI.

What is the symbol of phosphorus! Its equivalent?

Phosphorus. Equivalent 32. Symbol P.

Exp. 101. Wet the hands with water, and by means of the point of a knife, remove a stick of P from the bottle. Hold it between the thumb and finger in the air. It will smoke. P has combined with the O of the air, and formed PO^5 (phosphoric acid).

Explain Exp. 101.

With what has P combined?

The compound formed?

Exp. 102. Pour some water upon the table, and lay a stick of P in it. Cut off a piece of the size of a pea, and dry it between folds of blotting paper, taking care to avoid friction or pressure. Remove it quickly to another piece of dry blotting paper, and cover it with finely-pulverized

Explain Exp. 102.

C (carbon—charcoal). In a short time it will burst into spontaneous combustion.

What causes the combustion?

The C absorbs O from the air, and conveys it to the P so rapidly as to cause its ignition.

Exp. 103. Put a thin slice of P into a bottle, and pour over it an ounce of sulphuric ether. Let it remain four or five days, shaking it at least once each day. Pour off the liquid into another bottle. It

What inference is drawn from Exp. 103?

contains P in solution. Hence ether dissolves P.

Exp. 104. Drop some solution of P

Give Exp. 104.

upon the inside of the hand. It will emit a white vapor and an unpleasant smell. Rub the hand in the dark, and it will appear to be on fire. S (sulphur), C (car-

What have sulphur, phosphorus, and carbon been called? Why?

bon), and P (phosphorus) have been called *Pyrogens* (fire-producers).

Exp. 105. Fill a tall wine-glass two thirds full of boiling water; drop into it a small piece of P, and direct a stream of

If oxygen be brought in contact with phosphorus under boiling water, what follows? What will remain as a residue?

O upon it by means of the gas-bag and jet. A brilliant combustion will take place under water. A reddish mass will remain behind, which is P^3O (oxide of phosphorus).

How is phosphorus obtained?

P is obtained by decomposing bones, and must be preserved under water, as it undergoes oxidation in the air.

What is the equivalent of phosphuretted hydrogen? Its symbol?

Phosphuretted Hydrogen. Equivalent 35. Symbol PH³.

How is it obtained?

Exp. 106. Introduce an ounce of KO (caustic potash) into a half-pint glass retort, which fill ~~half~~ full of HO (water).

Explain the process in full.

Drop in a stick of P half an inch long, and place the retort in the sand-bath. A few minutes before applying heat, pour in a

Why was the ether added?

drachm of sulphuric ether, which, when heated, will volatilize and drive over the air that was contained in the upper part

What circumstance renders this precaution necessary?

of the retort. This precaution is necessary, as PH³ (phosphuretted hydrogen) explodes spontaneously in contact with air. The beak of the retort should pass into a

Fig. 38.

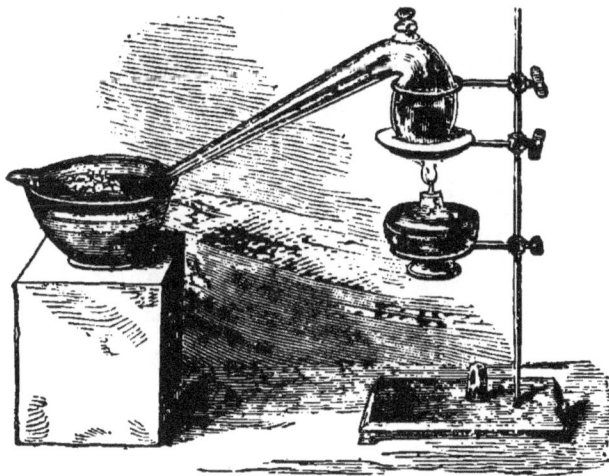

bowl filled with water, and must be immersed at least an inch and a half beneath its surface. The heat applied should be gentle at first, and increased until the gas passes over. Each bubble, as it rises to the surface of the water, bursts into combustion, and then forms a beautiful wreath of white smoke, which widens as it ascends, until it is finally dissipated in the air. PH^3 has an unpleasant odor.

As each bubble of the gas comes in contact with the air, what takes place?

What is the odor of PH^3?

Exp. 107. Hold a jar of O over the bubbles as they ascend through the water, keeping its mouth beneath the surface. Each bubble will emit a brilliant flash of light, and the vessel containing the O will be jarred.

If the bubbles of gas are exploded in oxygen, what is the phenomenon?

In the formation of PH^3, the water in the retort is decomposed. One atom of P unites with three atoms of H, and forms PH^3, while the three liberated atoms of O unite with another atom of P, forming PO^3 (phosphorous acid). Some authors affirm that two acids are formed together with PH^3, but the fact has never been established by experiment. The KO (potassa) undergoes no change, but simply acts upon the HO and P by its presence. Formula: $2P$; $3HO = PO^3$; PH^3. The P is thus divided between the elements of water.

What is decomposed when PH^3 is formed? Give the formula upon the blackboard.

What do some authors affirm?

Does the potassa undergo any change?

LESSON XXII.

Silicon. Equivalent 22. Symbol Si.

Si is the most abundant of all the elements except O. It is a brown solid, which burns vividly in O, but never occurs free in nature. It is always found combined with O, and is called Silex, Silica, or Silicic Acid.

Silica (Silicic Acid). Equivalent 46. Symbol SiO³.

SiO^3 abounds in all rocks except coal, limestone, and rock salt. It also enters

largely into the composition of soils. Sandstone and flint are nearly pure SiO^3 (sili-

ca). SiO^3, combined with different metallic oxides, forms the *rose quartz, chal-*

cedony, and opal. The stalks of grains and grasses owe their stiffness to SiO^3.

Scouring rushes possess it in much lar-

ger quantity. SiO^3, when subjected to a strong heat, possesses powerful acid prop-

erties. Heated with KO (potassa) or NaO (soda), it forms KO, SiO^3 (silicate of potassa), or NaO, SiO^3 (silicate of soda), ei-

Is common window-glass a salt?.

If sesquioxide of iron is present, what kind of glass is formed?

How is flint glass made?

ther of which compounds is glass ; hence common window glass is a salt. When Fe^2O^3 (sesquioxide of iron) is present, *green glass* is formed. If PbO (protoxide of lead), *flint glass.*

Exp. 108. Place half an ounce of KO (caustic potash), two ounces of HO (water), and one fourth of an ounce of pulverized sand, in a glass flask, and boil for half an hour, adding water as it evaporates. Transfer the whole to a tall phial, and allow the solution to cool and settle.

Give Exp. 108 in full.

Pour off the upper portion of the liquid into a larger vessel, and dilute it with six times its bulk of water, and add an ounce of HCl (hydrochloric acid). Now evaporate the solution to dryness in a glass or porcelain vessel. A white powder will remain, which is SiO^3 (silica or silicic acid). This powder can not again be dissolved in

In how many states does silica exist?

water. Thus we see that this body exists in two states, the soluble and the insoluble. In the soluble state, it is absorbed into the structure of plants, when the solution which contains it is evaporated by the sun's rays, and the SiO^3 (silica) is de-

What would follow if it were not for these two states?

posited. If it were now soluble, the rains and dews would dissolve it, and the plant would be left without support, as would be the human frame without bones.

What is the equivalent of fluorine? Its symbol?

Fluorine. Equivalent 19. *Symbol F.*

Has this element ever been isolated?

F has not yet been obtained uncombined with other elements, which is probably owing to its powerful affinities. At least this is the opinion of Professor Gregory, of Edinburgh. Its properties are, no

Why are its properties inferred to be similar to oxygen?

doubt, somewhat similar to O, as the two elements have no affinity for each other.

Give the equivalent of hydrofluoric acid.

Hydrofluoric Acid. Equivalent 20. *Symbol HF.*

Explain the process of obtaining HF.

Exp. 109. Place half an ounce of CaF (fluoride of calcium or fluor spar) in a shallow leaden vessel, which should be about three inches broad and three fourths of an inch deep. Add an ounce of SO^3, HO (sulphuric acid). Smear over a piece of plain glass two inches wide with wax. Form upon it, by removing the wax, letters or figures, and place them directly over the mouth of the vessel. (See Fig. 39.) Apply a gentle heat to the leaden vessel, and stand at a little distance for a few minutes. Remove the heat, and place the glass where it will cool. Care

Why should care be taken to avoid the contact of HF with the skin?

must be taken not to allow any of the fume of HF to come in contact with the skin, as it produces lingering sores. It must not be breathed. Wash the glass

Fig. 39.

with water; and remove the wax, when the letters or figures will remain distinct.

What is this process called? This process is called *etching on glass.*

Why must HF be preserved in leaden bottles? HF dissolves glass, and hence is preserved in leaden bottles. The materials used for obtaining HF were:

Materials used for obtaining HF?

Fig. 40.

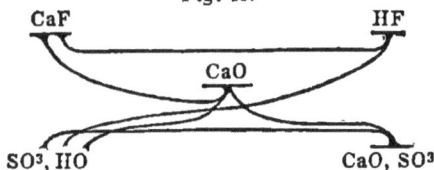

Draw and explain the figure.

The H of the acid was simply displaced by the metal Ca, when it united with the F, forming HF (hydrofluoric acid). Formula: CaF; SO^3, $HO = CaO$, SO^3; HF.

Give the formula.

What is the process when HF is desired in the liquid form? If HF is required in the liquid form, its vapor must be condensed in a receiver surrounded with ice, from which it may be poured into a leaden bottle and carefully sealed.

LESSON XXIII.

Aluminum. Equivalent 14. *Symbol Al.*

Exp. 110. Heat to redness in a deflagrating spoon (see Fig. 37) a small quantity of Al, and plunge it quickly into a jar of O, which must be kept covered. A brilliant combustion will ensue, the result

of which is Al^2O^3 (alumina or sesquioxide of aluminum), partially fused. It is near-

ly as hard as the diamond, glass being readily cut by it. Al^2O^3 is found nearly pure in nature, in the form of the *sapphire* and *ruby*. It is also a prominent ingredient of clays and slate rocks. Emery consists of minute particles of Al^2O^3, which, on account of their hardness, are used for polishing glass and the harder metals.

Exp. 111. Boil for twenty minutes half an ounce of Brazil wood with four ounces

of HO (water) in a retort or flask. Pour off the liquid, and add half an ounce of KO, SO^3, Al^2O^3, $3SO^3$ (alum). The color

E

is heightened. Now add some saturated solution of NaO, CO^2 (carbonate of soda). A brilliant red precipitate will settle to the bottom, which is *Brazil Wood Lake.* The *acetate* of *alumina*, formed by Exp. 8, is used by dyers for setting colors.

What is the precipitate called?
For what is the acetate used?

Sodium (Latin *Natrium*). *Equivalent* 23. *Symbol Na.*

What is the Latin for sodium?
Give its symbol and equivalent.

Exp. 112. Fill a test-tube half full of HO (water), into which drop a small piece of the metal Na. It will move round upon the surface, presenting the appearance of a beautiful silver ball, until finally it disappears. The element Na has united with the element O of the water, and formed NaO (soda).

If Na be placed upon water, what follows?

With what has Na united?
What compound is formed?

Exp. 113. Use boiling water instead of cold, and the combustion is violent, and often attended with explosion. The Na combines so vigorously with the O that a sufficient quantity of heat is liberated to ignite the escaping H, which, owing to the presence of Na, burns with a yellow flame.

How is the phenomenon changed by using boiling water?

What gas is ignited?
Why does it burn with a yellow flame?

Exp. 114. Pour the liquid formed by the above into an infusion of purple dahlia or blue cabbage. The color will be changed to green. All alkalies produce this effect upon vegetable blues; hence

How is soda proved to be an alkali?

NaO (soda) is an alkali. Na has a strong attraction for O, and must be preserved in a liquid which does not contain this element. Mineral naphtha is ordinarily used for this purpose.

How must sodium be preserved?

Chloride of Sodium (Common Salt).
Equivalent 58. Symbol NaCl.

What is the equivalent of chloride of sodium? Its common name? Its symbol?

Exp. 115. Fill a tumbler half full of Cl (chlorine) water, and add a piece of Na of the size of a pea. It will have the same appearance as in Exp. 112. But the Na has combined with the Cl instead of the O of the water, and formed NaCl (chloride of sodium). Evaporate the solution gradually, and cubical crystals will be formed. These are *Common Salt*.

Explain Exp. 115.

What are the cubical crystals formed?

Common salt is found in all parts of the earth in large quantities. Sometimes it is dug from the bowels of the earth in huge masses, which are afterward broken into small fragments, and thus rendered convenient for the use of man. In this form it resembles a transparent rock, and is called *Rock Salt*. At others it is obtained from *salt springs*, which are reached at a depth of several hundred feet beneath the surface of the earth. From these the water is pumped into vats and evaporated until it crystallizes. Granu-

In what parts of the earth is common salt found?

In what form is it obtained when dug from the earth?

What does it resemble? What is it called? Is it obtained in any other form?

How is granular salt obtained?
Why has the Creator furnished this substance in such abundance?

lar salt is formed by stirring the heated mass until it is nearly cold. Salt is indispensable to the life of plants and animals, and hence a benevolent Creator has furnished it in such abundance.

What is the Latin for potassium?
Its equivalent?
Symbol?

Potassium (Latin *Kalium*). *Equivalent* 39   *Symbol* K.

Exp. 116. Boil some water in a large basin, and when cold fill a test-tube from it. Invert the test-tube in the basin, and keep its mouth beneath the surface of the water. By means of a pair of slender pliers, place a piece of K (potassium) under the mouth of the tube, and allow it to escape.

If potassium be placed upon water, what follows?
What is liberated?

It will rise up in the tube, and combine rapidly with the O of the water in the form of a combustion, while the H is liberated, which will soon displace the water.

Exp. 117. Lift the test-tube from the

What is Exp. 117?

water, and quickly apply a lighted match or candle to its mouth. The H will ignite, and burn with its characteristic flame.

If potassium be brought in contact with ice, what ensues?

Exp. 118. Cut out a small cavity in a piece of ice, and drop into it a piece of K. A combustion will ensue. The K combines vigorously with the O of the ice (crystallized water), and the H is again

Give the formula.

liberated. Formula: HO; $K = KO$; H

Fig. 41.

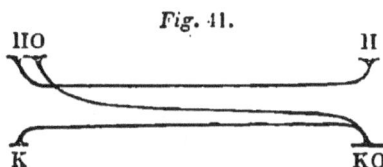

set free. K is a white metal resembling

What are the properties of potassium?

Na (sodium) in many of its properties. It yields to the pressure of the fingers like

What of its weight?

wax, and is the lightest of all the metals, being lighter than HO (water). It has a

How must it be preserved?

powerful attraction for O, and must be preserved, like Na, under naphtha.

LESSON XXIV.

What is the equivalent of manganese?
Its symbol?

Manganese. Equivalent 28. Symbol Mn.

Exp. 119. Mix one part of MnO^2 (peroxide of manganese) with one of C (charcoal) in a mortar, and pulverize them thoroughly together. Add sweet oil sufficient to form the whole into a thick

Explain Exp. 119.

paste. Transfer to a crucible, which should be covered and subjected to a white heat. The two atoms of O from the oxide unite with one of C, and form CO^2 (carbonic acid), and the Mn remains in its metallic form, slightly covered with an

Give the formula.

oxide of the metal. Formula: MnO^2; $C = CO^2$; Mn set free. (See Fig. 42.)

Draw and explain
Fig. 42.

Fig. 42.

MnO^2 CO^2

C   Mn

Exp. 120. Mix in a mortar one part of MnO^2, four of litharge, and four of pipe clay. Pulverize thoroughly, and add water until a thick paste is formed. Transfer the whole to a crucible, and apply a red heat. On cooling, it will form a bright black glaze. If half the quantity of MnO^2 is used, its color will be brown. Potters prepare black or brown glaze in a similar way. MnO^2 was formerly used instead of KO, ClO^5 (chlorate of potassa) for obtaining O.

How do potters prepare brown and black glaze? For what was the peroxide of manganese formerly used?

Mn is a gray metal, more difficult of fusion than Fe (iron). It is never found free in nature, but always in combination with O. It is obtained in great abundance in the form of MnO^2, from the mountains of Tennessee, and in many other parts of the world. It combines with O, forming several different compounds.

Give its properties.

Where is the oxide found?

Does manganese unite with oxygen to form more than one compound?

Magnesium. Equivalent 13. Symbol Mg.

What are the equivalent and symbol of magnesium?

This element is also a metal. It has a silvery appearance, and does not combine

In what form does this element exist?

Does it unite with O at ordinary temperatures? with O at ordinary temperatures. Hence it may be easily preserved.

The equivalent and symbol of magnesia?

Magnesia (*Protoxide of Magnesium*).
Equivalent 21. *Symbol* Mg O.

What of the diffusion of this compound? This compound is widely diffused throughout the earth, but is generally What is the composition of Epsom salts? combined with SO^3 (sulphuric acid), when it is called *Epsom Salts*. It is also found With what other body is magnesia found? in nature combined with CO^2 (carbonic acid). It is then called Carbonate of Mag- What do springs in various parts of the earth contain? nesia. Springs in various parts of the earth contain MgO, SO^3 (or SO^4, Mg) in solution, as at Saidschutz, in Bohemia, Where are large quantities of Epsom salts obtained? where large quantities of this salt are ob- tained. The waters are evaporated until What is the chemical name of Epsom salts? crystals of MgO, SO^3 (sulphate of magne- sia) appear.

Exp. 121. Dissolve as much Epsom salts in an ounce of water as possible, and Explain Exp. 121. add a strong solution of carbonate of po- tassa. A white precipitate will be form- ed, which is MgO, CO^2 (carbonate of mag-

Fig. 43.

Draw the diagram and explain it.

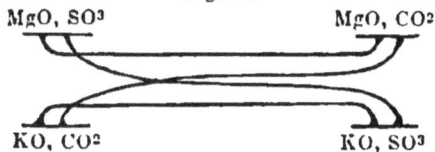

MgO, SO³   MgO, CO²

KO, CO²   KO, SO³

nesia). MgO, SO^3; KO, $CO^2 = KO$, SO^3; MgO, CO^2 set free. The SO^3 elects the

KO, and the CO^2 the MgO; that is, the acids have changed places—an illustration of *double elective affinity.* When

What is the compound of limestone and the carbonate of magnesia called? MgO, CO^2 (carbonate of magnesia) is found native in connection with CaO, CO^2 (limestone), the compound is called *Dolomite.*

Give the equivalent of calcium. Also its symbol. *Calcium. Equivalent* 20. *Symbol Ca.*

Have the properties of this element been fully investigated? The properties of this element have never been fully investigated, but it is Has it any affinity for O? supposed to be a metal. It has a powerful affinity for O.

What is the equivalent of protoxide of calcium? Its symbol? *Protoxide of Calcium* (*Lime*). *Equivalent* 28. *Symbol CaO.*

Exp. 122. Place a small piece of CaO, The common name for the compound? CO^2 (carbonate of lime—chalk) in a crucible, and subject it to a white heat for half an hour. Its properties are now Explain Exp. 122. changed. It will not mark, and has an alkaline taste. The heat has driven off the CO^2 (carbonic acid), and the CaO With what is lime usually found combined in nature? (lime) remains. CaO is usually found in nature combined with CO^2. Chalk, mar- What is the formula for gypsum and alabaster? Their chemical name? ble, and limestone are nearly pure CaO, CO^2 (carbonate of lime). Gypsum and alabaster are CaO, SO^3 (sulphate of lime).

Give the equivalent and symbol of chloride of calcium.

Chloride of Calcium. Equivalent 55. '

Symbol CaCl.

Explain the experiment for obtaining this compound.

Exp. 123. Add some pieces of chalk to an ounce of HCl (hydrochloric acid) until effervescence ceases. CaCl and HO

Fig. 44.

Draw and illustrate Fig. 44.

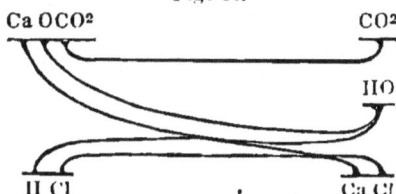

(water) are formed while the CO² (carbonic acid) is liberated in its gaseous form. Evaporate the solution until it has the appearance of a sirup, and allow it to cool. Crystals of CaCl will be formed.

For what is CaCl used? What name has been given to this salt?

CaCl is used chiefly for drying gases, as it attracts moisture with great force. Hence it is called a *hygroscopic salt.*

LESSON XXV.

What is the Latin for iron? Its symbol and equivalent? What is formed by evaporating the solution formed by Exp. 100?

Iron (Latin *Ferrum*). *Equivalent* 28.

Symbol Fe.

Exp. 124. Evaporate the solution formed by Exp. 100. Green vitriol will be formed, which is FeO, SO³ (sulphate of

Give the formula, the chemical and common name of the substance.

protoxide of iron), the substance used in

E 2

Exp. 4. It is commonly called Copperas.

What is said of iron? Fe is the most useful of all the metals,

Is it ever found nearly pure in nature?

With what elements is it usually found? and is sometimes found nearly pure in nature. It is usually found combined with O (oxygen), S (sulphur), or CO_2 (carbonic

Springs which contain the carbonate of iron are called what? acid). Springs which contain the carbonate of oxide of iron are called *Chalybeate.*

How many oxides are found in nature?

What are they? There are two oxides of Fe found in nature, the Fe^2O^3 (sesquioxide of iron) and the Fe^3O^4 (magnetic oxide—loadstone).

Is the protoxide ever found in nature? FeO (protoxide) is unknown in a free

What of its preparation in the laboratory? state in nature, and is prepared in the laboratory with great difficulty.

TABLE *of the Inorganic Elements, with their Equivalents and Symbols.*

	Equivalent.	Symbol.		Equivalent.	Symbol.
Selenium	40	Se.	Gold (Latin *Aurum*)	199	Au.
Bromine	78	Br.	Platinum	99	Pt.
Iodine	127	I.	Chromium	28	Cr.
Barium	69	Ba.	Antimony (Latin *Stibium*)	129	Sb.
Strontium	44	Sr.			
Cobalt	30	Co.	Arsenic	75	As.
Nickel	30	Ni.	Iridium	99	Ir.
Zinc	32	Zn.	Lanthanium	48	La.
Tin (Latin *Stannum*)	59	Sn.	Lithium	6	Li.
Cadmium	56	Cd.	Molybdenum	48	Mo.
Lead (Latin *Plumbum*)	104	Pb.	Osmium	100	Os.
Bismuth	71	Bi.	Palladium	53	Pd.
Copper (Lat. *Cuprum*)	32	Cu.	Rhodium	52	R.
Mercury (Latin *Hydrargyrum*)	203	Hg.	Ruthenium	52	Ru.
			Tantalum	185	Ta.
Silver (Latin *Argentum*)	108	Ag.	Tellurium	64	Te.
			Thorium	60	Th.

	Equiv- alent.	Symbol.		Equiv- alent.	Symbol.
Titanium	24	Ti.	Norium*	—	Nr.
Tungsten (Latin Wolfram)	100	W.	Didymium*	—	D.
			Glucinum*	—	G.
Uranium	217	U.	Niobium*	—	No.
Vanadium	69	V.	Ilmenium*	—	Il.
Yttrium	32	Y.	Erbium*	—	E.
Zirconium	34	Zr.	Donarium*	—	Do.
Terbium*	—	Tb.	Aridium*	—	Ar.
Pelopium*	—	Pc.	Cerium*	—	Cc.
			Boron	11	B.

Give the equiva-
lent and symbol
of selenium.

Selenium. Equivalent 40. *Symbol Se.,*

Its color.
Is it a solid?
What is its color
when in the form
of a fine powder?
For what ele-
ments has it an
affinity?

Se is a dark brown solid, having a metallic lustre and a deep red color when reduced to a fine powder. It has an affinity for O (oxygen), II (hydrogen), Br (bromine), Cl (chlorine), S (sulphur), and P (phosphorus).

What is the equiv-
alent of bromine?
Its symbol?
Is it a metal?

Bromine. Equivalent 40. *Symbol Br.*

What of its pois-
onous effects?

Br is a reddish brown liquid, very poisonous and corrosive to the skin. Three drops of it placed upon the tongue of a rabbit will produce death in a few seconds.

The symbol and
equivalent of io-
dine?
Give its proper-
ties.
In what form is it
obtained?
With what color-
ed flame does it
burn?

Iodine. Equivalent 127. *Symbol I.*

I is a bluish black solid, of a metallic lustre. It is obtained in thin laminæ or scales, and burns with a beautiful violet flame. It derives its name from the

* Elements whose equivalents are not well established.

From what does it derive its name?
When it unites with the metals, what are formed?
What is the equivalent of barium? Its symbol?

Greek *iodos*, which means *violet*. It unites with the metals, forming *iodides*.

Barium. Equivalent 69. Symbol Ba.

Its properties?

Ba is a gray metal, somewhat resembling cast iron. Its properties, however,

With what element does it unite?

are little known. It combines with O, forming BaO (protoxide of barium, or baryta).

What is the equivalent of strontium?
Give its symbol.

Strontium. Equivalent 44. Symbol Sr.

What of its properties?

Sr resembles Ba (barium) in appearance and properties, so far as it has been investigated.

Give the equivalent and symbol of cobalt.

Cobalt. Equivalent 30. Symbol Co.

Its color.

It is a reddish-colored metal, very diffi-

Other properties.

cult to fuse, and of a brittle texture. Co is strongly attracted by the magnet, and combines with O, Cl, and S.

What is the equivalent of nickel? Its symbol?
What metal does it resemble?

Nickel. Equivalent 30. Symbol Ni.

This element resembles silver in appearance. It is difficult to fuse, and very

For what is it sometimes used?

decidedly magnetic. Magnetic needles made of Ni are more durable than when made of steel, as they do not oxydize

What is the composition of German silver?

when exposed to air. Ni is one of the metals which form the so called German sil-

With what elements does nickel unite? ver, the other two being Cu (copper) and Zn (zinc). It unites with O, Cl, and S.

Give the equivalent and symbol of zinc. *Zinc. Equivalent* 32. *Symbol Zn.*

Zn is a bluish metal, not easily tarnished on exposure to air. At common tem- Its properties. peratures it is brittle, but when heated to 270° Fahrenheit it is both ductile and malleable. Burned in O, it emits a white light.

LESSON XXVI.

What is the Latin word for tin? Its symbol and equivalent? *Tin* (Latin *Stannum*). *Equivalent* 59. *Symbol Sn.*

Sn is a white metal having a decided lustre. It is but slightly oxydized on ex- Give its properties in full. posure to air, and is susceptible of being beaten into leaves not more than $\frac{1}{1000}$th of an inch in thickness. In this form it Of what is common tin-ware composed? is called *tin foil.* Thin sheets of iron coated with this metal form the common tin-ware of the shops. Sn in bars pro- With what bodies does the element unite? duces a crackling sound when bent. It combines with O, S, and Cl.

Give the equivalent and symbol of cadmium. *Cadmium. Equivalent* 56. *Symbol Cd.*

Cd resembles Sn (tin) in appearance,

but it is harder and much more tenacious.

Its properties. It is malleable and ductile to some extent, and, like Sn and Zn, unites with O, Cl, and S.

The equivalent of lead.
Its symbol.
Its Latin name. *Lead* (Latin *Plumbum*). *Equivalent* 104. *Symbol Pb.*

Give its properties.
How many oxides are formed? Pb has somewhat the appearance of Zn. It is soft, malleable, and ductile, but possesses little tenacity. When subjected to heat in air, two oxides are formed, PbO (protoxide of lead—a yellow powder) and Pb^3O^4 (red oxide of lead), which is formed when the heat is intense and the What is litharge? air is in excess. PbO, partially fused, forms the *Litharge* of commerce. The What forms the gray film upon the surface of melted lead?
With what other bodies does lead unite? gray film which accumulates upon the surface of melted Pb (lead) is PbO, mixed mechanically with the metal. Pb unites also with Cl, Br, I, and S.

Give the equivalent and symbol of bismuth. *Bismuth. Equivalent* 71. *Symbol Bi.*

Its color and texture.
At what temperature does it fuse?
What do several fusible alloys contain? This element is of a reddish color and crystalline texture. It fuses at a lower temperature than Pb (lead). Several fusible alloys contain it.

What is the Latin word for copper?
Give its symbol.
Its equivalent. *Copper* (Latin *Cuprum*). *Equivalent* 32. *Symbol Cu.*

Cu and Ti.(titanium) are the only red

For what is copper distinguished?

With what class of bodies will it readily unite?

What of the compounds formed? Why should cooking utensils made of copper be lined with tin?

What is formed when vinegar is allowed to stand in copper vessels? What is the composition of brass? Of bell-metal and bronze?

Give the Latin for mercury. What is its equivalent? Its symbol?

In what particular does it differ from all other metals?

What is it commonly called? What are its uses?

What compounds does it form with O?

What is the color of each?

What of its compounds with chlorine?

metals. Cu is distinguished for having the three properties, malleability, ductility, and tenacity. It will unite readily with any of the well-marked acids, forming compounds which are deadly poisons. Hence it should not be used for cooking utensils until it is lined with Sn (tin) or some other non-corrosive metal. Vinegar standing in copper vessels corrodes the metal, and the solution (acetate of copper) is highly poisonous. *Brass* is an alloy of Zn (zinc) and Cu. *Bronze* and *bell-metal* are composed of Cu, Zn, and Sn (tin).

Mercury (Latin *Hydrargyrum*). *Equivalent* 203. *Symbol Hg.*

Hg differs from all other metals in having the liquid form at common temperatures. It is commonly called quicksilver, and is used in barometers and thermometers. It is also used, when amalgamated with Sn (tin), for coating the back of mirrors. Hg unites with O, forming two distinct compounds, HgO (protoxide of mercury) and HgO^2 (peroxide). The former compound is a black, and the latter a red powder. *Red Precipitate* is HgO^2 (peroxide of mercury). This curious metal also forms two compounds with Cl, the $HgCl$ (protochloride of mercury—calomel) and $HgCl^2$ (bichloride—corrosive sublimate). It unites with I

With iodine? (iodine) in two proportions also, HgI (protoiodide of mercury) and HgI^2 (biniodide).

For what is the biniodide distinguished? This latter compound is distinguished for its brilliant vermilion color. (See Exp.

What of its compounds with bromine? 11, page 20.) Its compounds with Br (bromine) are similar to those of I.

What is the Latin word for silver? What is its symbol? Its equivalent? *Silver* (Latin *Argentum*). *Equivalent* 108. *Symbol Ag.*

The appearance of this metal is well At what temperature does it melt? known. It melts at a red heat, and, com- Combined with eight per cent. of copper, what does it form? bined with about eight per cent. of Cu (copper), forms our silver coin. It unites How many compounds does it form with O? with O, forming but one compound, AgO (protoxide of silver), which is a dark The oxide, when combined with ammonia, forms what? brown powder. This oxide, when combined with NH^3 (ammonia), forms a dangerously explosive body. Ag also unites With what other elements does silver unite? with S, Br, I, and Cl.

LESSON XXVII.

What is the Latin for gold? Give its equivalent. Also its symbol. *Gold* (Latin *Aurum*). *Equivalent* 199. *Symbol Au.*

What is said of its density? Au, with the exception of Pt (platinum), is the densest of all the metals, but Will it directly unite with O? it has feeble affinities. It will not direct-

Why will it not tarnish when exposed to the air?

ly unite with O, and hence its bright yellow color is never tarnished by exposure

How are oxides of the metal formed?

to air, water, or heat. Two oxides of the metal, however, are formed by an indirect process. It unites readily with Cl, form-

How many chlorides have been formed?

ing two compounds, which are AuCl (protochloride of gold) and $AuCl^3$ (perchloride).

What is the equivalent of platinum? Its symbol?

Platinum. Equivalent 99. *Symbol Pt.*

Its leading properties? How much heavier than water is gold? Platinum?

Pt is the heaviest of the metals. Au (gold) is about nineteen times heavier than water, and Pt is twenty-one times heavier than that body. It is the most ductile, and, with the exception of Cr (chromium), it is the most infusible. It

Has it ever been fused by the heat of the furnace?

resists the heat of the most powerful furnace, and hence is often used for making crucibles and roasting dishes. It can only

How can it be melted?

be melted by the oxyhydrogen blow-pipe, or by the agency of electricity. Pt re-

What does it resemble in appearance?

sembles silver in appearance, and com-

Will it unite directly with O?

bines with O, S, and I, but, like Au (gold), it will not unite directly with O.

What are the equivalent and symbol of chromium?

Chromium. Equivalent 28. *Symbol Cr.*

Its properties? Has it ever been fused or acted upon by acids?

This element exists in the form of a gray metal. It has never yet been fused nor corroded by the strongest acids, in which respects it differs from all other

Has it been adapted for any practical use?

metals. As yet, it has been adapted to no practical use.

What is the Latin for antimony?
Its symbol and equivalent?

Antimony (Latin *Stibium*). *Equivalent* 129. *Symbol Sb.*

Its properties?

Sb is a brittle, though valuable metal.

What are printing-types composed of?

It has a bluish-gray color, and is the chief element of printing-types. It combines

How many compounds does it form with O?

with O, forming four distinct compounds, two of which are oxides, the others being acids. Sb^2O^3 (sesquioxide of antimony)

What is the active principle of tartar emetic?

is a gray powder, and is the active principle of *tartar emetic*, that compound being a double tartrate of antimony and potassa

What is said of the peroxide?

$\left(\begin{matrix} Sb^2O^3 \\ KO \end{matrix} \right\} C^4H^2O^5 \right)$. SbO^3 is a heavy powder, resembling the preceding in properties, with the exception that it is a more deadly poison. SbO^4 (antimonious acid)

Give the other compounds.

is a white powder, insoluble in water, and is very infusible. It unites with bases, forming salts called *Antimonites*. SbO^5 is nearly similar in properties to SbO^4.

Give the equivalent and symbol of arsenic.

Arsenic. Equivalent 75. *Symbol As.*

Its chief properties.
How is it distinguished from all other metals?

This metal has a light gray color, is of brittle texture, and when burned emits the smell of garlic. This distinguishes it from all other metals. If entirely pure, its lustre is not tarnished by exposure to air, unless it is strongly heated.

Exp. 125. Place six or eight grains of As (metallic arsenic) in the centre of a glass tube, and apply heat by means of the spirit-lamp. The smell of garlic will soon be perceived (which should not be breathed to any considerable extent), and the metal will soon be dispersed over the upper part of the tube in the form of a beautiful black mirror. Hence As volatilizes by heat.

Arsenious Acid. Equivalent 99. *Symbol As O^3.*

AsO^3 is commonly called *Rat's Bane* or *Arsenic.* It is a heavy white powder, having no smell, and but very little taste. It is a deadly poison, and is often used for criminal purposes. Its antidote is Fe^2O^3, HO (iron rust), or the white of eggs, the latter of which should be administered freely.

Exp. 126. Take a glass tube one foot in length, and heat one end of it in the flame of a spirit-lamp; draw it to a point, which hold in the flame until the orifice is closed by fusion. When cool, place it in an upright position, and introduce two

Give Exp. 125 in full.

Does this metal volatilize by heat?

Give the equivalent and symbol of arsenious acid.

What is it commonly called? Its properties?

For what purposes has it been used? What are its antidotes?

Give Exp. 126 in full.

or three grains of AsO^3 (arsenious acid). Drop into the tube a slender piece of charcoal. Hold it horizontally in the flame until the coal glows, and quickly transfer the heat to the end of the tube which contains the AsO^3, which will be volatilized, and, passing over the glowing coal, will release its O^3, and the As will form a metallic mirror upon the sides of

Place the formula illustrative of the change upon the black-board.

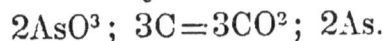

the tube just above the coal. Formula: $2AsO^3$; $3C = 3CO^2$; $2As$.

Give the equivalent and symbol of B.

Boron. Equivalent 11. Symbol B.

What are its chief properties ?
Where is it obtained ?
How combined ?

This element exists in the form of an olive-colored solid. It is obtained from the hot springs of Italy, combined with

In this form what is it called ?

O. In this form it is called boracic acid (BO^3). But it is difficult to ascertain whether this compound is really an acid

How does it effect the tests for acids and alkalies ?

or an alkali. It browns turmeric paper like an alkali, and reddens litmus like an

Does it combine with alkalies ?

acid. It however combines with NaO (soda) and some other alkalies.

For what elements has it no affinity ?

The element B has no affinity for H, I, Br, or Cl.

The following is a list of rare and generally unimportant elements, whose properties are not well known: Iridium, Lanthanium, Lithium, Molybdenum, Osmium, Palladium, Rhodium, Ruthenium, Tantalum, Tellurium, Thorium, Titanium, Tungsten, Uranium, Vanadium, Yttrium, Zirconium, Terbium, Pelopium, Norium, Didymium, Glucinum, Niobium, Ilmenium, Erbium, Donarium, Aridium, and Cerium.

LESSON XXVIII.

Acids.—Bases.—Salts.

What are acids?

ACIDS are bodies which usually have a sour taste, and change vegetable blues to red. Some acids, however, do not possess these properties. They all have the property of neutralizing alkalies and other bases.

Do all acids possess these properties?
What property do they all possess?

Bases are bodies which have an attraction for acids, and, when alkaline, change vegetable blues to green, also red test-paper to blue. Acids and bases unite and form a numerous class of salts, according to the commonly-received theory upon this subject. The more simple theory, by which all the phenomena of the formation of salts are explained, is to consider the subject in the light of *radicals* and *metals* instead of *acids* and *bases*. Example : Cl and Na unite and form common salt, which was the original type of all the salts. As Cl is regarded as a simple body, we shall consider it as a *simple salt radical*, which, when brought in contact with the metal Na, the salt is formed. Some-

What are bases?

What effect have alkalies on vegetable blues?

What do acids and bases form?

What more simple theory is given?

Give the example.
What does Cl and Na stand for?
What was the original type of all the salts?
How is Cl regarded?
What is formed when chlorine and sodium are brought in contact?

Is the salt radical ever a compound body? Give an example.

times the salt radical is a compound. Example : Cy (NC^2, cyanogen) is a compound body, and is a salt radical; Cy and the metal K unite and form the salt KCy

What is said of nitric acid?

(cyanide of potassium). Again: NO^5, HO (nitric acid) is now regarded as a hydrogen acid, having the formula NO^6, H.

What takes place when this acid combines with a metal?

Hence, when nitric acid combines with a metal and a salt results, the hydrogen of the acid is simply displaced by the metal.

Give the formula illustrating the decomposition.

NO^6, H ; K = K, NO^6 ; H. Here the H is liberated, and the K (potassium), united with the compound radical NO^6, forms the salt, whose formula is usually given

What is the formula of the salt? Is sulphuric acid now regarded as an oxygen acid? Why is it considered a hydrogen acid?

as KO, NO^5. SO^3, HO (sulphuric acid) is now generally admitted to be a hydrogen acid, as it possesses no acid properties without the presence of this element. SO^3 was formerly regarded as an oxygen acid, and SO^3, HO as a hydrate of this

If it is a hydrogen acid, what should be its formula?

acid; but, regarding it as a hydrogen acid, we shall have the formula SO^4, H ; Fe =

If iron be brought in contact with SO^4, H, what follows?

Fe, SO^4 ; H. Here the metal Fe (iron) displaces the H, and the salt is formed.

Why do some authors object to this view as the basis of a system?

Some American authors object to this simple view as the basis of a system, because SO^4 has never been isolated; but it will be remembered that the same objec-

How do they regard NO^5?

tion will apply to the compound NO^5, which, the same authors contend, exists.

Are both these bodies hypothetical?

Is there any well-marked acid that does not contain hydrogen?

What is said of dry SO^3, PO^5, and CrO^3?

Has CO^2 the power of neutralizing the alkalies?

Do those compounds which were formerly called oxygen acids possess acid properties when hydrogen is not present?

What is Professor Gregory's definition of a salt?

Both these bodies are hypothetical. It may be added that there is no well-marked acid that does not contain H (hydrogen). Dry SO^3, PO^5, CrO^3, and several other oxygen compounds, have no acid properties. CO^2 (carbonic acid), though commonly called an acid, has not the power of neutralizing the alkalies. Indeed, none of those compounds which were formerly called oxygen acids possess positive acid properties without the presence of hydrogen.

Professor Gregory, of Edinburgh, gives the following definition of a salt: "It is the compound formed by replacing the hydrogen of an acid by a metal."

Sulphates.

When salts contain two equivalents of the acid or radical, what prefix is used?

Sometimes salts contain two equivalents of the acid or radical, when they have the prefix bi- ; as, bisulphate of potassa. Neutral sulphate of potassa has the formula K, SO^4, or, according to the old theory, KO, SO^3.

Exp. 127. To a saturated solution of carbonate of potassa add sulphuric acid till effervescence ceases. Carbonate of potassa $=$ KO, CO^2, or K, CO^3. Sulphuric acid $=$ SO^3, HO, or SO^4, H. K, CO^3 ; SO^4, H $=$ K, SO^4 ; CO^2 ; HO. K, SO^4 (sul-

Give Exp. 127 in full.

The formula illustrative of the change.

How does sulphate of potassa crystallize?
phate of potassa) crystallizes in six-sided prisms, which contain no water.

What is the formula for sulphate of iron?
Fe, SO^4, or FeO, SO^3 (sulphate of iron). The common name of this salt is *Copper-*

For what is this salt used?
as. It is much used by ink-makers and dyers. (See Exp. 99, page 88.)

What is the formula of Glauber's salt? In what form are its crystals?
Na, SO^4, or NaO, SO^3 (Glauber's salt), crystallizes like K, SO^4, but the prisms are much larger, and contain ten atoms of water of crystallization. •

Give the formula of sulphate of baryta.
Ba, SO^4, or BaO, SO^3 (sulphate of baryta), usually occurs in nature as large

Is it soluble in the acids? Is it soluble in water?
tabular crystals. This salt is not soluble in the acids nor in water. It has no water of crystallization.

What is the formula for sulphate of lime? Has this salt more than one form? Where is it found?
Ca, SO^4, 2HO, or CaO, SO^3, 2HO (sulphate of lime). This salt has a variety of forms. It is found native in the Mammoth Cave, Kentucky, and in many other

What are selenite, gypsum, and alabaster?
parts of the world. Selenite, gypsum, and alabaster are different forms of it.

How is plaster of Paris obtained?
Plaster of Paris is obtained by depriving the salt of its water.

Give the formula of sulphate of alumina. Will this salt crystallize?
Al^2, 3SO^4, or Al^{2}O^3, 3SO^3 (sulphate of alumina), has never yet been made to crystallize until it is combined with some other salt. It is one of the salts which unite to form *alum*, the other being K, SO^4 (sulphate of potassa). Alum, then,

What is the composition of alum?
is a double salt, composed of sulphate of

alumina and sulphate of potassa. (See Exp. 8, page 18.)

Give the formula of sulphate of magnesia.

Mg, SO^4, HO, or MgO, SO^3, HO (sulphate of magnesia), is commonly called

What is its common name?

Epsom salts. It may be readily formed

How may this salt be formed?

by dissolving carbonate of magnesia in dilute sulphuric acid. Its crystals are four-

The shape of its crystals?

sided prisms.

What general observation is made in reference to the sulphates?

The sulphates are a numerous family of salts, but most of them do not occur in nature, and as yet are of little use.

LESSON XXIX.

Nitrates.

How may the nitrates be formed?

THE nitrates, like the sulphates, may be obtained by the action of NO^6, H, or NO^5, HO (nitric acid), on the metals or

Are nitrates insoluble in water?

metallic oxides. All nitrates are soluble

What effect is produced on these salts when they are exposed to a red heat?

in water, and are decomposed at a red heat. The most important salt of this

Which is the most important salt of this family?

family is the K, NO^6, or KO, NO^5 (nitrate

What is its common name?

of potassa). Its common name is *Salt-petre.* It is found abundantly in nature

Where is it found?

in crystals, but most commonly mixed

Is it usually found pure?

with soil, called *Nitre-beds.* It is the

Of what is it the chief ingredient?

chief ingredient of gunpowder (see Exp. 70, page 71), and, mixed with sulphur and

F

How is fulminating powder formed? carbonate of potassa, a compound called *Fulminating Powder* is formed.

Exp. 128. Mix thoroughly in a mortar six parts of K, NO^6 (nitrate of potassa);

Give Exp. 128. four of K, CO^3, or KO, CO^2 (carbonate of potassa), and two of sulphur. Place a

If a grain of the mixture be immersed in the flame of a spirit-lamp, what follows? grain of the mixture upon a slip of cop- per, and immerse it in the flame of the spirit-lamp. A loud report will take place.

What is the formula of nitrate of soda? Na, NO^6, or NaO, NO^5 (nitrate of soda), is found native in the East Indies and in

What of its properties? Peru. Its properties are very similar to K, NO^6, only that it burns more slowly when mixed with charcoal.

What is the formula of nitrate of ammonia? NH^3N, O^5, or $O^5N^2H^3$ (nitrate of am- monia), has already been described on page 63.

Give the formula of nitrate of baryta. For what is it chiefly used? When exposed to a red heat, what result takes place? Ba, NO^6, or BaO, NO^5 (nitrate of bary- ta), is chiefly used as a chemical test, and when exposed to a red heat, the Ba re- tains one atom of O, and NO^5 is driven off in the form of N ; O^5.

What is the formula of nitrate of strontium? For what is it used? What color does it impart to flame? Give the formula of nitrate of copper? What is the form and color of this salt? What does it yield when heated to redness? Sr, NO^6, or SrO, NO^5 (nitrate of stron- tium). This salt is used extensively in the manufacture of fire-works. It im- parts a brilliant crimson flame.

Cu, NO^6, $3HO$, or CuO, NO^5, $3HO$ (ni- trate of copper), is formed in deep blue crystals, which, when heated to redness, yield protoxide of copper.

What is the composition of nitrate of mercury? The composition of nitrate of mercury is not well established.

The formula of nitrate of silver? For what is it used? Ag, NO⁶, or AgO, NO⁵ (nitrate of silver, or lunar caustic), is used to eschar the skin and to destroy tumors. It is

Of what is it the active principle? What takes place with all the compounds of silver when exposed to light? How does this salt crystallize? When fused and run into moulds, what is it called? also the active principle of indelible ink. All compounds of silver are blackened when exposed to light in contact with organic substances. This salt crystallizes in thin tables, and, when fused and run into moulds, is called *lunar caustic.*

Chlorates.

This class of salts is similar to the nitrates, but the only important ones are Which are the important salts connected with the chlorates? chlorate of potassa and chlorate of baryta.

Give the symbols of chlorate of potassa. How does it crystallize? How much water is required to dissolve it? At what temperature does it fuse? If the temperature is increased, what follows? K, ClO⁶, or KO, ClO⁵ (chlorate of potassa). It crystallizes in six and four sided tables, and is soluble in sixteen times its weight of water. It fuses at 500° Fahrenheit, and, when the temperature is increased, pure O is liberated. See page 41.

Exp. 129. Place in a mortar two grains of sulphur and six of K, ClO⁶ (chlorate of potassa); pulverize them thoroughly together with a pressure not exceeding ten What is Exp. 129? pounds. Collect the whole into a conical pile, upon a smooth stone or other hard surface, and strike the mass with a hammer. A deafening report will follow.

Give Exp. 130 in full.

Exp. 130. Cover a piece of P (phosphorus), of the size of a radish-seed, with pulverized K, ClO^6, and strike the mass forcibly, as in Exp. 129. Another loud report will ensue. K, ClO^6 is one of the ac-

Of what is chlorate of potassa an active principle?
How may it be decomposed?

tive principles of percussion powder, also of lucifer matches. It is decomposed by some of the stronger acids.

Exp. 131. Fill a wine-glass with hot water, in which place five grains of P and ten of K, ClO^6. Now bring in contact

Give Exp. 131.

with the mass, by means of the dropping-tube, some strong sulphuric acid. The P will burn under water. The salt is decomposed, and its O liberated, which produces the combustion.

It was once attempted to use this salt

What was the result of attempting to use this salt in the formation of gunpowder?

instead of nitre in the formation of gunpowder, but, on pulverizing the mass, it exploded, spreading destruction far and wide.

Exp. 132. Place ten drops of HCl (hy-

What is Exp. 132?

drochloric acid) and ten grains of this salt in a pint of rain water. The solution has

What is the formula of chlorate of baryta?
The form of its crystals?
In how much cold water is it soluble?
For what is this salt sometimes used?

marked bleaching properties.

Ba, ClO^6, or BaO, ClO^5 (chlorate of baryta), crystallizes in four-sided prisms, and is soluble in about four times its weight of cold water. This salt is sometimes used for obtaining ClO^4 (chlorous acid).

LESSON XXX.

Phosphates.

What is the term commonly applied to the formula PO^5?

PO^5 (page 46) is called phosphoric acid, as this is the term commonly applied to this compound. It will be remembered,

Does this compound possess acid properties?

however, that PO^6 possesses no acid properties until it has combined with an atom

When acting as an acid, what is its probable composition?

of HO, when its probable composition is PO^6, II. Authors generally take the po-

What position do authors generally take in reference to the different compounds of P, O, and II?

sition that there are different hydrates of PO^5, but as these hydrates require a differ-

Do these hydrates require the same or a different amount of base?

ent amount of base, it would seem more natural to consider each as a distinct

Give the formula of the so called hydrates.

acid. The so called hydrates are PO^5, HO; PO^5, $2HO$; and PO^5, $3HO$. The

What is the probable composition of the acids?

acids are probably PO^6, H; PO^7, H^2; and PO^8, H^3. It will be remembered that all

How do all hydrogen acids form salts?

hydrogen acids form salts by replacing the hydrogen with a metal. If there be but

If there be but one atom of hydrogen in the acid, how much of the metal will be required to form the salt?

one atom of hydrogen in the acid, but one atom of metal will be required to form a

What is understood by a neutral salt?

neutral salt, that is, a salt which has neither acid nor alkaline properties. If the acid contains two or three atoms of hydrogen, two or three atoms of the metal will

What is the acid called which has but one atom of hydrogen?
If it contains two atoms, what?
If three atoms? be required to form the salt. An acid containing one atom of hydrogen is called monobasic, one containing two atoms bibasic, and one containing three atoms tribasic. PO^6H is a monobasic acid, PO^7H^2 a bibasic acid, and PO^8H^3 a tribasic acid. Then we shall have of the phosphates of soda,

Give the monobasic, bibasic, and tribasic phosphates of soda.

Monobasic Acid.	Bibasic Acid.	Tribasic Acid.
Na, PO^6.	$2Na, PO^7$.	$3Na, PO^8$.

Who first suggested this theory? Liebig first suggested this theory, which has since been gradually gaining favor with the progressive chemists of the age, How is it now regarded? until it is regarded as an essential part of the science.

Chromates.

Give the formula of bichromate of potassa.
What are its color and properties?
For what is it used? $KO, 2CrO^3$ (bichromate of potassa) is a beautiful red crystalline salt, which is easily soluble in water, and is used extensively in calico printing. Its solutions What effect do its solutions produce upon the skin? should not be brought in contact with the skin, otherwise lingering sores will be produced.

What is the formula of chromate of lead?
Its properties? PbO, CrO^3 (chromate of lead) is a powder which is not soluble in water, and has a fine yellow color. It is commonly call- What is it commonly called? ed *Chrome Yellow.* (See Exp. 9, page Give the formula of dichromate of lead. 19.)

$2PbO, CrO^3$ (dichromate of lead) is

commonly called *Red Lead*, and is thus found in nature. Its form is crystalline, and it is used extensively as a paint.

Borates.

NaO, $2BO^3$ (biborate of soda—*borax*) is the only important salt of the borates. It is used as a flux for welding and soldering, on account of its solvent power when heated to redness.

Carbonates.

CO^2 (carbonic acid) and BO^3 (boracic acid) ought not really to be called acids, as neither has power to neutralize the alkalies. Even two equivalents of CO^2 or BO^3 to one of KO (potassa), form alkaline salts. *E. g.:* KO, $2CO^2$ (bicarbonate of potassa) possesses nearly as strong alkaline properties as the KO did before it was united with the $2CO^2$.

KO, CO^2 (carbonate of potassa). This compound will not crystallize, and is used in mineral analysis as a flux.

KO, $2CO^2 + 2$ aq.* (bicarbonate of potassa). This body crystallizes, and does not deliquesce† on exposure to air.

* Aq. means water, and when used in connection with salts, it denotes the water of crystallization.

† Deliquesce means to melt or turn to liquid. Ef-

Side notes (left margin):

What is it commonly called?

Where is it found, and what is its use?

What is the formula of biborate of soda? Of what is it the only important salt? What are its important uses? Why?

Should CO^2 and BO^3 be called acids?

Why not?

Will two equivalents of either of these bodies neutralize the alkaline properties of potassa?

Give the formula of carbonate of potassa. For what is it used?

What is the formula of bicarbonate of potassa? Has this salt any water of crystallization? Does it deliquesce when exposed to air? What is the meaning of deliquesce?

$Na, CO^3 + 10$ aq., or $NaO, CO^2 + 10$ aq.
(carbonate of soda), is a body which crystallizes, and, when exposed to air, undergoes efflorescence. Soap and glass mak-
ers use it in large quantities, as it an-
swers nearly the same purpose as KO (potassa), and is much cheaper.

Exp. 133. Add an ounce of Na, CO^3
(carbonate of soda) to a washing-tub full
of hard water. It will be rendered soft.

$Na, 2CO^3 + $ aq. (bicarbonate of soda).
This is another alkaline salt, which exists
in the form of a white powder. It forms
the effervescing property of Seidlitz pow-
ders.

$2NH^3, 3CO^2, 2HO$ (sesquicarbonate of
ammonia) is a hard, crystalline salt,
which gives off the strong smell of ammonia. It effloresces in the air by losing
ammonia. The powder is a $NH^3, 2CO^2$
(bicarbonate).

CaO, CO^2 (carbonate of lime) exists in
nature in the form of chalk, limestone,
and calcareous spar. Oyster and other
shells are chiefly composed of it. Spring
water, when passing over limestone, dissolves small portions of it, and if brought

floresce means to pass into a fine powder or dust.
Effervesce signifies to bubble; effervescence is a bubbling.

in contact with more CO^2 (carbonic acid), the CaO, CO^2 is dissolved, which renders the water hard. (See Exp. 133, page 128). The soda unites with the excess of CO^2, and the CaO, CO^2 is precipitated, when the water is rendered soft.

Why will soda render hard water soft?

Pb, CO^3, or PbO, CO^2 (carbonate of lead), is the common white lead of the shops. It is sometimes found in nature, when it is called White-lead Spar. It is the most deadly of all the poisonous compounds of lead. If pure water be allowed to stand in leaden vessels exposed to air, it will in a short time contain small particles of the PbO, CO^2. These particles often prove destructive to health, though not taken in sufficient quantity to produce death.

What is the formula of carbonate of lead?
What is its common name?
When found in nature what is it called?
What of its poisonous properties?
Should water be allowed to stand in leaden vessels?

Hence pure water conducted through leaden pipes should not be taken into the stomach. Should the water contain portions of other salts before coming in contact with the lead, it is less dangerous to use it. There are many other salts, which are unimportant to the student, which have been omitted for the sake of brevity.

Should pure water which is conducted through leaden pipes be taken into the stomach?
If the water contained salts before coming in contact with the lead, is it equally dangerous to use it?

F 2

LESSON XXXI.

Fats and Alkalies.

What are proximate principles ?

COMPOUNDS which exist in plants and animals, ready formed, are called *proximate principles.* Stearine and oleine, the chief ingredients of animal fats, are proximate principles. , These two bodies may be regarded as salts, stearine being a stearate of oxide of glyceryle, and oleine an oleate of the same base. The oxide of glyceryle is the sweet principle of all animal fats and oils. When the base KO (potassa) or NaO (soda) is brought in contact with stearine and oleine, the weaker base oxide of glyceryle is liberated by single elective affinity, and the two *fat acids,* oleic and stearic, unite with the potassa or soda, and form stearate and oleate of these bases, either of which is *soap.*

What are the chief ingredients of animal fats? Are they proximate principles?

How may they be regarded ?

What is the sweet principle of animal fats and oils?

What becomes of the glyceryle when potassa or soda is brought in contact with stearine and oleine ?

With what do stearic and oleic acids unite ? What do they form ? Give the common name of the compound.

Soft Soap.

Describe the experiment for making soft soap.

Exp. 134. To a drachm of caustic potassa add an ounce of water ; transfer to a chemical flask ; add an ounce of mutton tallow, and boil for half an hour. Pour off the liquid into a larger vessel, and add soft water gradually until the mass assumes a jelly-like appearance. It is soft soap.

How is hard soap formed?
What alkali is used in forming soft soap?
In forming hard soap?

Exp. 135. Use caustic soda in the above instead of potassa, and *hard soap* will be formed. Hence soft soap is a potassa soap, and hard soap a soda soap.

How may soda soap be obtained from soft soap?

Exp. 136. Dissolve some soft soap in hot water, and add some table salt. The soap will rise to the surface in a condensed mass. It is now soda soap. This process

Why is this process resorted to by soap-makers?

is resorted to by soap-makers on account of its cheapness when compared with hard soap formed directly from caustic soda.

From what is Marseilles soap made?

Marseilles soap is made from soda and olive oil. When colored with metallic oxides and perfumed, it is called *Castile*

What do soaps contain which impart strong lathers?

soap. Soaps which impart thick lathers contain cocoa-nut oil as an ingredient.

Bread-making.

What are the proximate principles of flour?

Bread can be made from flour, which contains starch, sugar, and vegetable fibrin. These compounds are proximate

What is vegetable fibrin sometimes called?
What is first done with the flour?

principles, and the latter is sometimes called gluten. The flour is first mixed with water (dough), when yeast or leaven is incorporated with it, and it is allowed to stand in a warm place until the mass increases considerably in bulk. It is then

What next ! Do the properties of the flour undergo a change by being baked ? In the process of fermentation, what gas is set free? In what form ? What do these bubbles cause?

subjected to heat (baked), and the properties of the flour undergo a decided change. In the process of fermentation (rising), CO^2 (carbonic acid) is liberated in small bubbles, which cause the increased size of the loaf, as well as its porous appearance when baked. This pro-

This process is called what?
What proximate principle of the flour is acted on by the yeast?
Into what is it converted ?
Is the starch changed during fermentation?
Does any part of the fibrin disappear with the sugar?
Why is it desirable to save the sugar and fibrin?
How may the object be accomplished ?

cess is called the *vinous ferment.* It is the sugar of the flour that is acted on by the yeast, which is converted into carbonic acid and alcohol, both of which escape in baking. The starch is unchanged during fermentation, but the sugar and a part of the fibrin disappear. The fibrin and sugar being nutritive, it is desirable to save them. In order to accomplish this object, bread is raised by means of carbonate of soda and diluted hydrochloric acid. The soda is first dissolved in water and incorporated with the flour,

Give the process.

after which the diluted acid is quickly kneaded into it. The stronger acid decomposes the base, soda, and the carbonic acid is liberated.

The formula.

Formula : NaO, CO^2; $HCl = HO$; $NaCl$; CO^2. It will be seen

Name the resulting compounds.

that the resulting compounds are water, common salt, and carbonic acid. By this mode the fibrin and sugar are saved, together with that portion of water which is not driven off in the form of vapor. Bread contains about one sixth part of its weight of water in the solid form.

How much water, in the solid form, does bread contain?

As the student of this neglected branch of popular education has now crossed its threshold, it is to be hoped he will not content himself with having mastered the few elementary principles embodied in the foregoing pages, as the greater novelties and beauties of the subject are yet to be unfolded. By becoming still more familiar with the invisible causes which produce the prominent phenomena of chemical science, he will be irresistibly led to behold nature in the light of a vast chemical laboratory, performing upon a grander scale the same processes that the chemist so successfully imitates in his limited sphere. The rain, the dews, the snow, the hail, the breeze, the hurricane, the water-spout, the earthquake, and the volcano, are all phenomena which result from those chemical laws which produce, by their varied action, all that is beautiful, terrible, or sublime in nature.

APPENDIX.

Solution of Acetate (Sugar) of Lead.

To 2 oz. of water add a quarter of an oz. of acetate of lead. Allow the mixture to stand one hour, during which time it should be frequently stirred with a glass rod. Pour off the liquid and filter it.

Solution of Sulphate of Iron (Copperas).

Add half an oz. of sulphate of iron to half a pint of water. Stir the solution frequently, and, after having stood an hour, it should be filtered.

Infusion of Nutgall.

Pulverize a light-colored nutgall in a non-metallic mortar, and add 4 oz. of water. Stir with a glass rod, and at the end of 15 minutes filter the liquid.

Solution of Oxalic Acid.

Add a drachm of oxalic acid to 2 oz. of water.

Solution of Potassa.

To 2 oz. of water add a quarter of an oz. of potassa.*

* All chemical solutions should be preserved in ground-stoppered bottles.

Solution of Sulphate of Copper (Blue Vitriol).

Add a quarter of an oz. of sulphate of copper to 2 oz. of water. Allow the solution to stand one hour, and filter it. A few drops of this liquid to a test-tube half full of water forms the solution used in Experiment 5.

Solution of Nitrate of Mercury.

To a quarter of an oz. of water and half an oz. of nitric acid add a drachm of mercury. Allow the liquid to remain 8 hours, and pour it off (decant).

Solution of Alum.

To 2 oz. of water add half an ounce of alum. Allow the solution to stand 2 or 3 hours, and transfer the liquid portion.

Solution of Bichromate of Potassa.

To 2 oz. of water add 4 grs. of bichromate of potassa.

Solution of Iodide of Potassium.

To 2 oz. of water add 8 grs. of iodide of potassium.

Solution of Bichloride of Mercury.

To 2 oz. of water add 8 grs. of bichloride of mercury (corrosive sublimate). Allow the liquid to remain 3 hours, and then filter it.

Infusion of Blue, or Purple Dahlia.

Pour warm water upon the petals of this flower until they are covered with the liquid. Allow it to stand in a warm place for several hours, when it may be filtered.

Solution of Nitrate of Silver.

To 2 oz. of water add a drachm of nitrate of silver (lunar caustic).

Solution of Chloride of Calcium.

Add pieces of chalk or marble to 2 oz. of hydrochloric acid until effervescence ceases. Transfer the liquid portion.

Litmus Paper.

Place 2 drachms of litmus in a flask, and pour over it 2 oz. of water. Heat the mixture nearly to boiling for one hour, and, when cold, filter it. Pass strips of white unglazed paper through the blue liquid until, on drying, they assume a decidedly blue color. To the remaining solution add lemon-juice gradually until it assumes a red color. Slips of paper may now be drawn through this until they are red when dry. The first is called blue test-paper, and the last red.

THE END.